GREEN ENERGY SOURCES AND FUTURE ENERGY SCENARIO

Deepika: Many belonging to upper middle class and upper think that they can afford to pay for electricity at high rates and therefore, there is no need for them to cut down the consumption of electricity.

Mukesh: That is correct. There is another side to this issue. Many do not pay for the electricity used by them, or rather wasted by them. This is the case with those working in public and private institutions. There is no ownership for the problems arising out of wasteful consumption by the employees of these. Very often managements do not put in efforts to educate their employees on the related aspects, and thereby avoid the avoidable wastage of electricity. The slogan, energy saved is energy produced, must percolate down to all sections of our society, encompassing administrators and the administered in an equal measure.

Anuradha: There seems to be uniformity of opinion among energy administrators on this issue. When opinions differ, the general public is confused.

Deepalakshmi: Considering the need for effecting energy conservation at all costs, there could be a need for downward revision of [illegible] requirements for various categories of [illegible]. In addition, methods could be devised for lighting only as need. In [illegible], it is worthwhile to recall the procedure Indian railways [illegible]. They have implemented a novel method of controlling the lighting of railway platforms, only for a predetermined duration, in relation to the presence of train at the platform.

Deepa: I think that [illegible] and administrators [illegible] to pay more attention to this area of energy savings, [illegible] that another method [illegible] for managing the whole issue. It is [illegible]

[illegible] is probably information on the

GREEN ENERGY SOURCES AND FUTURE ENERGY SCENARIO

S. Muthukrishna Iyer

KALPAZ

ISBN : 978-93-5128-000-2
Price : ₹ 650

First Published 2014

Published by

Kalpaz Publications
C-30, Satyawati Nagar,
Delhi-110052
Ph. : 9212142040
E-mail: kalpaz@hotmail.com

Laser Typesetting : Rudra Computer Graphics, Delhi
Printed at : G. Print Process, Delhi

Cataloging in Publication Data—DK
Courtesy: D.K. Agencies (P) Ltd. <docinfo@dkagencies.com>

Muthukrishna Iyer, S., 1947-
Tourism and Public Relations/Annamalai Murugan.
p. cm.
Includes index.
ISBN 9789351280002

1. Green technology. 2. Renewable energy sources. 3. Energy development. I. Title.

DDC 628 23

Dedication

This book is dedicated to those unfortunate homosapiens who are destined to live in the energy starved era; we seem to be at its door step.

Contd. Table III

Year	Population (in Crore)	Rate of growth at 10 year interval (%)
2061	203.6	7.0
2071	213.8	5.0
2081	221.6	4.0
2091	[illegible]	[illegible]
2101	228.2	0.0

The world population which reached 700 Crore in October 2011, is expected to be around 1010 Crore at around 2050, and 1170 Crore at 2100. This data is from the book titled "Global Energy Perspectives," Edited by Nebojsa Nakicenovic, Arnulf Grubler and Alan McDonald and published by Cambridge University Press in the year 1998.

It is predicted that the increase in world population in the coming decades will largely be in the less developed countries of the world, and India is one among them.

2. Population and energy utilisation are closely related

Human population that existed at any period of time and the energy that was available then for use are related. This is clear from the fact that our population touched one billion (100 crore) only around 1804 A.D. The population also used to dip in situations of war and natural calamities.

The mining, processing and utilisation of hydrocarbons in the form of coal and petroleum on a large scale, has made the world population to increase in very large numbers during the twentieth century and later, in comparison to that existed at any point of time in the history of mankind. In addition, the availability of this additional quantum of energy as mentioned above, has enabled the maintenance of large portion of human population at a higher level of comfort compared to the past.

Contents

installed capacity could grow to this maximum value and then continue to be available at the same level.

4.2.1.2. Electricity generation from non-green energy sources

The difference between the demand for electricity and the extent of its availability from green energy sources is to be met from thermoelectricity generation. At a level of demand which is 2 times the current level, the installed capacity of thermoelectric plants is to be around 252,200 MW (this is twice the present capacity). Assuming that the demand at 2100 is 3 times the present level, the installed capacity of thermoelectric plants need be around 118,800 MW (this is around 3.6 times the present capacity).

The following points may be noted with respect to thermoelectricity generation:

- The thermoelectricity generating capacities in demand will be lower to the extent of energy intensity improvements achieved with respect to these plants. Units being established presently could be providing superior performance levels compared to the ones already in operation.

- Large scale capacity additions of coal based thermoelectric plants is to be brought in only by closing our eyes to global warming related issues.

- Large scale use of thermoelectric plants working on gaseous petroleum could address the environment related issues partially.

Considering the above factors we could conclude that the availability of electricity from coal based and natural gas based thermoelectric plants will be limited by the extent of availability of these fuels from time to time.

The above analysis clearly indicates that the role of green energy sources in the area of electricity generation in our country

Foreword

"Energy can neither be created nor destroyed; it can only be converted from one form to another"; this is a law of thermodynamics. Prior to the industrial revolution, the limited natural resources could satisfy the energy needs of the people. In contrast, today, our lifestyle has changed beyond recognition and fuels power almost every activity of our lives. Dwindling reserves of conventional energy resources and the ever increasing demand for energy has been an area of concern for the researchers and planners alike for more than two decades. Today, this has become a matter of grave concerns and the entire population has to be sensitized about the need for conserving energy and resorting to alternate and renewable sources like solar, wind, geothermal, ocean waves etc.

It is in this context that the book titled *Green Energy Sources and Future Energy Scenario* by Shri S. Muthukrishna Iyer, becomes relevant. Shri S. Muthukrishna Iyer, a retired ISRO engineer, provides a complete exposition of the world energy scenario with special reference to India. He makes a detailed study on the possible role of the new green energy sources in meeting our future energy needs by analyzing the "net energy" available from each of these sources. The distinction between the high energy intensity of combustion used for generating electricity and the low energy intensity obtained from weak energy sources such as solar rays, ocean waves/tides and geothermal heat has been clearly brought out. He has carried out a comparative analysis on the viability of energy conversion processes for electricity generation from different sources like solar ray, ocean waves, geothermal energy, nuclear etc. Shri S. Muthukrishna Iyer also stresses the need for energy - population - economy planning from a long term perspective.

The language is simple and lucid with attention in minor details, and suggestions provided throw insights into the benefits and problems in the adoption of these renewable sources. On the whole, I am sure this book is very useful for the researchers and the common man alike. I wish him all success.

M.C. Dathan
(Distinguished Scientist/Director)
Liquid Propulsim Systems Centre (LPSC)
Indian Space Research Organisation (ISRO)
Thiruvananthapuram

Preface

The shortages that are likely to come in the availability of fuels will have adverse implications on economies; particularly in a highly populated country like ours. For the same reason, the responsibility of providing thoughtful and reasoned out information on availability of energy in the coming times lies with the present day energy professionals and administrators.

The goal of writing this book is to highlight the imminent shortages that are likely to happen in the area of availability of energy in the coming decades and centuries, its repurcussions our economy and in turn on the human race.

This book is the outcome of an analytical study on the possible contribution of green energy sources on the future energy scenario and contains a whole lot of new inferences.

Thiruvananthapuram **S. Muthukrishna Iyer**

It is well known that liquid petroleum has been in use for over a century; however, the use of natural gas in our country is comparatively of recent origin. Under these circumstances, shortages if any experienced in the availability of liquid petroleum during the period upto 2050 may be compensated by the use of natural gas to the extent the same is technically feasible.

In addition, the demand for liquid petroleum is expected to be on the rise during the coming decades and the new discoveries could compensate for the same. However, it is possible that the availability of liquid petroleum declines any time after 2050 and the same could bring in very serious repurcussions on the sustainability of our economy.

The above data also indicates that we Indians constituting around 17 per cent of the world population as of now, is blessed with only around 0.42 per cent of world liquid petroleum reserves.

4.2.2.3 Gaseous petroleum

(1) Data on natural gas reserves is given below:

Country	*Proven reserves (trillion cubic metre)*	*Production (trillion cubic metre)*
Russia	47.57	
Iran	33.6	
Turkmenistan	26.2	
Qatar	25.47	
U.S.A.	7.0	
Saudi Arabia	8.2	
Venezuela	5.524	
Nigeria	5.246	
Algeria	4.504	
China	3.100	
Sub total	187	
World Total	190	3.359

(The above data is taken from internet)

Acknowledgements

My Daughter Lekshmi had been a souce of inspiration for writing this book.

My father in law Shri SS Ayer (93 years) read through the manuscript, and suggested improvements in language used also pointed out certain errors and omissions. I express my sincere gratitude to them.

Introduction

The field of energy has been fascinating me for long; we live by utilising it in some form or another.

Sun is our prime source of energy. Other than the uninterrupted availability of energy from sun, many factors are responsible for evolution of life forms on earth. The precise positioning of earth in our solar system is one.

The evolution of plants and their producing seeds in quantities very much more than that is required for the continuation of their species, are among the reasons for animal kingdom to exist on earth. We humans with freedom to move, often migrate to greener pastures, looking for congenial, comfortable and biologically rich environments. However, thanks to our recent advances in space science and technology, some of us think that in future we would leave greener pastures and migrate to settlements in barren ones like the Moon or Mars.

Solar energy causes photosynthesis in plants; also causes rains. Rain water nourishes life forms on earth. By utilising sun's heat our ancestors dried agricultural products and preserved them. They were aware of sun's role in the creation and maintenance of life forms on earth.

In the recent past, we discovered coal and petroleum in the earth's crust. The invention of coal fired steam engines created demand for coal and thus, it became a key energy source of the times. Liquid petroleum and natural gas were discovered later. Then onwards, coal, petroleum derivatives and natural gas played their roles in fuelling our economy.

Coal and petroleum are highly concentrated energy sources formed in the earth's crust over millions of years of weathering on biological forms. It may be noted that barring the power of flowing water used for hydro electricity generation and the energy intensity available in nuclear reactors, there are no other energy forms on earth, with the level of energy intensity comparable to that of fire, obtained by burning fuels.

The large scale use of coal triggered the first industrial revolution. Later on, liquid petroleum was mined, refined and distributed far and wide. The wide spread use of these enabled us to come up with chains of inventions and their innovative uses mechanised most of our routine activities.

Notable among these are automobiles we brought into existence more than a century ago. For the wide spread movement of automobiles, networks of railway lines and roads were built. Mechanised ships moved goods in bulk far and wide. This revolution in powered transportation enabled large scale distribution of coal, petroleum and agricultural/industrial goods. This also resulted in the improvement of living standards of our population in many parts of the globe and in turn, became a catalyst for the increasing the rate of rise of human population.

The invention of aircrafts, a large variety of communication systems and their extensive applications made the world one global village enabling us to reach far off locations in less than a day, and also telecommunicating instantly. We should also recall the progress made in metallurgy as well as energy intensive manufacturing processes. Without these, the growth in the manufacturing sector we are seeing today would not have been possible.

The large scale availability of coal and petroleum energized space transportation too. Step by step, our probes moved out from orbits of earth to the Moon and planets of our solar system.

Of course, such changes have their side effects. When the use of fuels grew in an exponential manner, the content of carbon dioxide in our atmosphere also went up. This is believed to be causing warming of our environment on a global level, and is feared to result in sea level rise, and in turn submerge many populated cities across continents. In addition, severe losses in agricultural production are feared. The signs of global warming are said to be visible in the melting of Arctic ice blocks, at rates higher than what we experienced in the past.

Alarmed about the possibility of calamities we expect in a few decades from now, we are looking for reducing the emission of carbon dioxide to substantially lower levels by cutting down the consumption of fuels.

However, the suggestion to reduce the consumption of fuels is easier said than done. It simply could mean that we reduce the consumption of energy, and face the consequences of severely curtailed economic activity. This is not a feasible suggestion, since, our lifestyle as it is today is enabled primarily by the utilisation of hydrocarbons mined out and distributed liberally.

The long-term alternative is believed to be the development of energy sources belonging to environment friendly category, and using them for meeting our energy needs. By adopting this route, it is envisaged that the consumption of hydrocarbons could be reduced. The environment friendly category of energy sources are known as green energy sources.

Production of electricity from hydro, wind, ocean waves/ tides, geothermal energy, ocean thermal energy, nuclear fission and fusion and direct solar energy are considered as promising in the coming centuries. Another proposal under active consideration is the change over from hydrocarbon economy to hydrogen economy.

However, we cannot assume that the issue of availability of energy from environment friendly energy sources is solved by simply allotting quotas for them. Hydroelectricity and wind electricity are proven sources we can depend upon for meeting our energy needs to the extent of their potentials. These energy sources are found to be viable, provided, the power production units are installed at locations having specified mean wind speed for wind electric units, and an acceptable level of water head for hydroelectric units.

Fission based nuclear electricity generation is used to varying degrees in many countries in a viable manner. However, safety related issues are coming to fore and causing uncertainties of serious nature on its future. These issues seem to be serious to the extent that they might retard or altogether stop the developmental activities of new types of fission based nuclear reactors. We know that developing workable units of Fast Breeder Reactors and bringing them into stream on an industrial scale is essential to ensure the availability of nuclear fuels for longer durations, however, this is one area wherein the present turmoil might adversely affect fission based nuclear electricity generation programme.

Creating mini suns on earth in the form of nuclear fusion reactors and deriving energy from them in a large scale is a big dream of ours. However, the process of their creation is extremely complicated and therefore, we cannot be sure of these units providing us energy in the foreseeable future.

Other green energy sources referred above are being explored by different agencies primarily for electricity generation for the past many decades. The possibility of substitution of these for conventional energy sources will essentially not depend on the availability of some quantum of electricity from them, but on proving their viability for electricity generation. The issue of viability of various green energy sources is an extremely important aspect for their adoption for electricity generation on a large scale.

It is believed that the viability related analysis presented here shall trigger further in depth studies in this area.

"Green energy sources and future energy scenario" is chosen as the title of the book. This is felt appropriate for the following reasons:

- The results of the analysis on the viability of green energy sources already in use and the new ones we hope to come to our rescue in the coming times, has relevance both for the immediate future as well as beyond.
- Mankind lived on earth for long by deriving energy from renewable energy sources in the form of plant residues, solar energy, wind and flowing water. The human population that existed varied from time to time, depending on the extent of energy available at their disposal.
- The large scale utilisation of non renewable energy forms in the form of coal and petroleum is a very recent phenomenon. By pumping in energy from these on a large scale, we have created an economy that is many orders voluminous compared to the one that existed in the pre-coal/petroleum era.
- The utilisation of hydrocarbons in a large scale is believed to have brought in man made global warming, the related ill effects and therefore, we believe that their consumption need be brought down over the coming decades and centuries.
- The solution lies in moving in to an economy that is fuelled largely by green energy sources.
- The general belief seems to be that green energy sources such as solar energy, geothermal energy, ocean wave/tide energy, energy from nuclear fusion and hydrogen are sure energy sources for the future, and we will be able to harness energy from these, in forms and quantities as demanded from time to time. However, we have not yet found ways

and means to overcome the technological hurdles that might stand on our way of their utilisation in a viable manner.

- It is in this context that the relevance of this study comes in. Our objective is to segregate the green energy sources that are likely to be viable sources for generating energy in forms we desire, from others that are prima facie non-viable.

This author wishes that this study will be continued by the energy planners and administrators in the national/international level.

Targeted readers:

- Energy providers shall be immensely benefited by this study. It is important for them, to pursue with activities beneficial to the humanity as a whole. It is also essential for them, to pursue with 'real goals', and look at energy related issues that could arise beyond a few generations.

- Energy administrators play an important role in the planning and financing part. There seems to be wide variation in perceptions on many factors related to energy among those working in the field, however, the reality could be only one. The future of energy availability as well as the future of mankind are interrelated. Therefore, it is essential that energy administrators move way from perceptions to reality. Conclusions drawn in this book are expected to help them in this regard.

- Scientists and technocrats are expected to contribute by carrying out detailed studies on the various aspects related to harnessing energy and its utilisation in a viable manner. The future of energy as well as the future of the mankind depends on a realistic evaluation of the issues involved in the energy area by our scientists and technocrats.

- Students who aspire to be the agents for providing energy for the future generations.

- The general public whose existence depends on the availability of energy as needed.

In short, since energy is everybody's business, the book is sure to benefit all those who decide to take the benefit of its contents.

1

Sources of Energy

We use energy as heat as well as electricity. We shall begin with understanding the energy sources used for electricity generation.

1. Established Energy Sources for Electricity Generation

1.1. *Thermoelectricity*

From the beginning of electricity generation, coal and bio fuels fuelled thermoelectric power plants.The use of some petroleum fractions came in later; and the use of natural gas is picking up. We believe that there are no limits to capacity additions in thermoelectricity generation area. We are also in the process of augmenting its installed capacity at a feverish pace.

However, the following issues are believed to be adverse to deriving electricity from thermo electric plants to the extent we want to:

- The issues related to global warming demand that we bring down the consumption of coal as years pass by.

- We are facing many hurdles wih respect making coal available to the power stations as required.

- Environmental issues other than global warming come up with large scale establishment of thermoelectricity generation facilities.

1.2. *Hydroelectricity*

We have been using the power of flowing water for generation of electricity for over a century, by establishing hydroelectric power plants. This belongs to green energy category; the cost of electricity generated is the lowest too. We have exploited its potential to a significant extent, and presently, we meet around 20 per cent of the electricity needs of our country from these. However, in our country, there does not look to be scope for significant quantum of capacity additions in this area in the coming times, since, setting up of new plants is being shot down in the name of environmental protection.

1.3. *Fission based nuclear electricity*

Having brought into the arena of electricity generation after 1950's, many advanced countries established reactors to the extent that they met a significant portion of their needs from nuclear electric plants. Many countries desire to create substantial generating capacities in this category. However, some are viewed with the suspicion, since, it is persumed that their interest is on making atom bombs. Our country is keen to go in for accelerated capacity additions through import of high capacity units, however, many issues are standing on our way of achieving the goal.

Reserves of nuclear fuels are finite and hence, this energy source belongs to non-renewable energy category. It is also to be noted that, nowadays, conflicting type of reports are coming out regarding the manner in which countries such as Germany and Japan intend to handle the nuclear electricity generation scenario of the coming times.

1.4. *Wind electricity*

Wind electricity generation is comparatively a new entrant into the area of power generation. Units are established at locations with favorable wind characteristics. Wind electricity generating units established at sites after ensuring the availability of quality wind for significant part of a year, function in a viable manner.

These are the energy sources we utilise for setting up electricity generating plants as of now.

2. New Green Energy Sources Considered as Promising for Electricity Generation

2.1. *Solar energy*

We use solar energy for providing electricity in remote areas on earth and powering satellites. The use of solar electricity was confined to special areas primarily due to the high cost of plants, and their unwieldy nature. However, in India, currently we are promoting the establishment of grid connected solar electricity generation in a big way. Being a green energy source and also since solar energy is believed to be abundantly available at most locations in our country, it is perceived by some among us that establishment of such units will augment to our electricity generation capacities significantly. The supporters of wide spread use of solar electricity generation also hope that in the long run, the price of solar electricity could be brought down matching to the cost of electricity generated from conventional energy sources.

However, many factors are standing on the way of generating solar electricity in a viable manner.

2.2. *Geothermal energy*

This vast heat source which is available beneath the earth's surface is perceived as promising for electricity generation. Experimental units are set up in many countries utilising the thermal energy derived from geothermal wells. In general, the intensity of heat energy available from this source is very low except for the isolated cases of availability of quality steam in large quantities trapped underground at some locations. Some countries seem to be having favourable locations for utilising this energy source in a viable manner largely in the form of heat energy, particularly in the vicinity of vulcanoes.

2.3. *Ocean waves/tides*

Energy contained in ocean waves/tides is another source of energy that is eternal. Experimental units belonging to these categories are already setup and are under evaluation. In the same lines as solar and geothermal energy sources, the energy intensity contained in these is low and for the same reason, before adoption of this technology for commercial exploitation, viability aspects of such units need be carefully evaluated.

2.4. *Nuclear fusion*

Sun generates thermal energy by fusing light nuclei such as hydrogen into heavy ones. We believe that by creating mini suns on earth, we shall be able to generate limitless quantum of energy from fusion based nuclear reactors. However, the technologies involved are extremely complicated and therefore, we are yet to make a breakthrough.

The above aspects are discussed in the following chapters in an evaluative manner.

3. Fuels/other Thermal Energy Sources

Coal

Coal is used to fuel thermoelectric power stations and industrial processes. In countries with extreme cold climate, environmental conditioning inside buildings consumes a significant quantum of energy and coal is used for this purpose.

Before petroleum became available, coal was used for production of liquid and gaseous fuels. It is envisaged that we shall establish coal liquefaction/ gasification plants in a big way in future when reserves of petroleum deplete. This is, since, the availability of coal is considered to be significantly more than that of petroleum.

Thermoelectric plants emit carbon dioxide and is said to cause global warming. This perception of ours has resulted in demand for bringing down the quantum of coal burned in thermal stations/industries.

Petroleum

Liquid petroleum

Liquid petroleum was discovered by the middle of the nineteenth century. The development of technologies related to its mining, processing and that of various transportation media matured and came into wide spread use largely in the twentieth century. Eventhough concerns are raised at times on possible depletion of its reserves, we continue to mine out and use liquid petroleum according to our wishes. According to a report published in *The Hindu* of January 17, 2012, India's import content stands at around 75 per cent. We are utilising around 134.76 million tonnes of liquid petroleum annually.

Natural gas

In our country, large scale use of natural gas is comparatively a recent phenomenon. The transportation modes for the same are transnational pipelines and ships provided with liquefaction facilities. Natural gas is expected to replace liquid petroleum distillates in some of its applications. This is a cleaner fuel compared to coal as well as liquid petroleum. Since natural gas was brought into use comparatively in the recent times, it is assumed that its reserves will outlast that of liquid petroleum.

Currently, we are facing problems in making natural gas available for fuelling our economy to the extent of its demand. Even political issues are hindrances for laying transnational pipelines which are considered to be the convenient mode for its transportation.

Extent of availability of the above mentioned fuels.

It is appropriate to look into the extent of availability of coal and petroleum for our future use. In order to have an understanding on the same, the relevant data as published in the book named *Global Energy Perspectives,* published by Cambridge University Press (1998) is quoted here.

Note:

- Reserves are those occurrences that are known, and are recoverable with the present day technologies and market conditions.

- Resources are occurrences that are distinct from reserves in that they have less certain geological assurance, or lack present economic feasibility.

Table: Data on Fuels (giga ton oil equivalent)

Fuel	*Utilization during the year 1990*	*Reserves*	*Resources*	*Additional occurrences*
Crude oil				
Conventional	3.2	150	145	
Non-conventional				
Natural gas	-	193	332	1900
Conventional	1.7	141	279	400
Non-conventional				
Hydrates				
Coal	-	192	258	18700
Uranium	2.2	606	2794	3000
For Fast Breeder Reactors	0.5	57	203	150
	-	3390	12150	8900

- Resource base is the sum of reserves and resources, and consists of the following:

  Potentially recoverable coal, conventional oil and natural gas.

  Unconventional oil resources such as oil shale, tar sands and heavy crude.

  Unconventional natural gas resources such as gas in Devonian shale, tight sand formations, geo pressurized acquifiers and coal seams.

- Additional occurrences are those which are considered as not recoverable and are excluded from the resource base of the future.

- Typically, unconventional sources defer from conventional sources by one or more of the following characteristics:

  They occur in significantly lower concentrations.

  They require unusual or extreme technological prerequisites for their recovery.

  They need complex and capital intensive conversion processes for bringing them into usable form.

  They have significant environmental implications.

We have the history of predictions of the past on the availability of petroleum going wrong a few times and this indicates the difficulty encountered in estimating their overall availability. The table given above contains many figures that are primarily relevant and less relevant for our estimation, hence it could be difficult for any one to get a real picture of the energy scenario of the coming times.

Table: Availability of Fuels in giga ton oil equivalent

Fuel	*Utilization during the year 1990*	*Reserves*
Crude oil (conventional)	3.2	150
Natural gas (conventional)	1.7	141
Coal	2.2	606
Uranium	0.5	57
Fuel for Fast Breeder Reactor	-	3390

The relevant portion is presented in the above table:

The table indicates the availability of crude as 150 giga ton oil equivalent and its use during the year 1990 was 3.2 giga ton oil equivalent. Assuming that we used 5 giga ton oil equivalent during the year 2010, the reserves of 150 giga ton oil equivalent could last only for a period of three decades. This indicates an alarming situation we need to seriously debate.

Bio fuels

We are looking for the production of bio fuels in a big way for the following reasons:

- Switch over to the use of these in a large scale could reduce our dependence on petroleum.
- Bio fuels are eternal sources of green energy.
- We are hoping that that there are many plants whose cultivation in a big way could result in large scale bio fuel production.
- Brazil has been able to meet a large portion of their needs for liquid fuels in the form of ethanol extracted from sugar cane.

However, except for the production of ethanol from sugar cane, we might encounter viability related issues. Moreover, shortage in the availability of land dedicated to the cultivation for bio fuels could be a bottleneck in most countries.

Solar Energy

Solar energy is considered to be a promising thermal energy source for the coming times, particularly for heating water meant for various applications. Considering the enormous quantum of fuel/electricity being used for providing hot water in cold countries, solar water heaters are promising.

Fusion Based Nuclear Energy

A very high degree of uncertainty exists with respect to the development of fusion based nuclear power reactors. However, in case we are able to succeed in bringing them to wide use, thermal energy derived from them could be of use then.

Geothermal Energy

Geothermal energy available beneath the surface could be used largely for district heating purposes, particularly in cold countries.

4. Change over to Hydrogen Economy

A big step aimed at reducing the level of atmospheric pollution is hoped to be the adoption of hydrogen economy. Production of hydrogen in a very large scale, and the development of related energy conversion devices such as fuel cells and hydrogen storage devices for use in mobiles, are prerequisites for the adoption of hydrogen economy. However, eventhough there was high level of enthusiasm during the initial phases of proposing the adoption of hydrogen economy, the present day scenario seems to be that many hurdles are standing on our way of the development of related technologies.

2

Electricity Generation (Non-green) – Thermoelectric Power Plants

1. Thermoelectricity Generation is Non-green

We largely utilise hydrocarbon based fuels for fuelling thermo-electricity generating plants. Fuels belong to non-green energy category.

2. Functioning of Thermoelectric Plants

In thermoelectric plants, turbines are run by the energy of superheated steam produced in boilers fired by coal, naphtha, liquid petroleum gas or fuels of biological origin. Steam expands during its passage through the turbine blades, and in the process, a significant portion of its kinetic energy is transferred to the turbine shaft in the form of rotational force, and this in turn powers the electricity generator.

3. Energy Intensity Available from Fuels

The energy intensity available from the co nbustion of fuels is the highest from among the energy sources used of for the generation of electricity.

4. Issues Related to Bringing Down the Consumption of Coal

In spite of this high energy intensity available from fuels, we desire to cut down their use, since, we have the problem of excessive quantities of carbon dioxide being let out into atmosphere. However, we shall not be able to move away from them for the following reason:

Energy is utilised for the growth and maintenance of the world economy. Many of the economically advanced nations are in the forefront with respect to their per capita income, which means that they are burning fuels in large quantities and thereby are providing better living standards for their countrymen. They believe that for maintaining their life style, economy need be maintained at least at its current levels and therefore, the energy input into their economy is not likely to come down significantly over the coming times. Additional demand for energy in the coming decades might be compensated by energy intensity improvements.

At the same time, we have a large population in developing countries waiting for being provided with improved living conditions. In our country, we have unacceptable level of poverty in many regions and the dire need to take remedial actions. In addition, our population is expected to grow at high rates althrough the present century. Therefore, it is essential that our economy keeps growing at matching rates for a long time to come. Maintaining higher growth rates demand proportionately higher quantum of energy to be pumped into our economy.

In such a situation, when we are expected to curtail the consumption of coal and petroleum derivatives by a significant extent, a major portion of our energy needs of the coming times has to necessarily be met from green energy sources; this is our general belief. It means that the utilisation of green energy sources has to be augmented manyfold in comparison to its present levels. This aspect is being discussed in detail in the following chapters.

5. Conclusions

* The energy intensity available from the burning fuels is the highest. This makes thermoelectricity generation viable.

* Reserves of coal are believed to outlast the reserves of liquid petroleum. Hence, coal could fuel thermoelectric plants for many more decades after the reserves of liquid petroleum are depleted.

* The utilisation of fuels in a big way for the past few centuries has enabled the progress in science and technology to today's level.

* According to study reports, the consumption of fuels at 2050 is expected to be two to three times the present level, particularly in developing countries like ours.

* At such high levels of consumption, the liquid petroleum is projected to last only for few more decades. Experts often give warnings in these lines.

* We are concerned about the use of fuels at high rates, and the environmental damages caused by them.

3

Proven Green Energy Sources

We have been using green energy sources for generation of electricity for some time now. The oldest among them is the hydro power. Production of fission based nuclear energy came in later, and many technologically and economically advanced countries use it for electricity generation. Some countries are believed to be using the thermal energy from these reactors to heat up buildings. Harnessing electricity from wind is comparatively a recent phenomenon. In the following chapters, we shall discuss the various aspects of harnessing electricity from these proven green energy sources.

Hydroelectricity

1. Hydroelectric potential

1.1. Hydroelectric potential is green and eternal

Hydroelectric potential is an eternal green energy source. This aspect clearly indicates the need for utilising this energy source to the maximum extent at the earliest.

1.2. Origin of hydroelectric potential

Solar energy falling on the surface of the oceans and fresh water collections generates water vapour, and this accumulates as clouds. It takes weeks or even months for the clouds to precipitate and rain down. A small portion of the rain water is stored in the reservoirs of hydroelectric plants The potential energy of this water is used to power hydroelectric units.

All these make hydroelectric plants the most favoured and most economical ones for the production of electricity. The country with abundant hydroelectric resource is blessed and benefiting the most out of the advancements made in the fields of science and technology during the past couple of centuries.

1.3. Kerala – a land of rivers

Our state is a narrow strip on the Western side of Western Ghats. We have 44 small rivers originating from the mountain ranges, and flowing towards the West. Power plants are established utilising water from these. Probably, we could say that hydroelectric potential is one of the very few natural resources the State of Kerala is blessed with. This is extremely valuable due to the fact that our region lacks with respect to the availability of many minerals our neighbouring states are blessed with in abundance.

The hydroelectric power units we have largely belong to medium and small head categories. Until a couple of decades ago, our electricity generation was mainly from these. However, it is worth noting that we have been able to build only around 1500

Table

Serial No.	*Project*	*Projected Installed Capacity (MW)*
1.	Athirappally	163
2.	Patrakkadavu	70
3.	Kuriarkutty-Karappara	84
4.	Twin kallar	60
5.	Pooyamkutty	210
	Total	**587**

MW installed capacity in this area, and currently we are facing severe objections with respect to building new ones. The State of Kerala has a long list of hydroelectric projects waiting for nod from the authorities for construction. A list of these, as published in the *Malayala Manorama* daily of March 29, 2012 is given below:

The above list points out to the ground reality, that eventhough we have more locations wherein we can build hydroelectric power stations, our hands are tied. The hope for establishing more hydroelectric power stations is receding, particularly after the report of Madahv Gadgil committee has been submitted.

Any delay in utilising this eternal green energy source for electricity generation will only help to add to global warming woes.

1.4. Hydroelectric potential varies among regions

Flowing water had been a source of power for electricity generation for more than a century. There are a few countries in the world with significant hydroelectric potential, some others have it to moderate levels and there are also some with no hydroelectric potential at all.

We have an example for a country with little or no hydroelectric potential in the Kingdom of Netherlands in the North Western Europe. The country is of plain land with an average height of 11 meters above sea level (ref: *Malayala Manorama Year Book* – 2007). Much of the land is below sea level, and is protected by dykes stretching to a length of 2,400 km.

1.5. What limits hydroelectric potential

Hydroelectric stations are established in mountainous regions, wherein, rain water could be collected in huge quantities and utilised for the generation of mechanical power. The availability of such favourable locations provide hydroelectric potential. Locations of reservoirs have to be such that the stored water could be brought down to lower levels through open canals, or pipelines. When water falls through a height, its potential energy is converted into kinetic energy and this powers hydroelectric power production units.

2. Hydroelectricity Generation

Rain water falling on raised land has potential energy relative to oceans towards which it flows. This energy is converted into mechanical power by passing the stored water through a hydraulic prime mover and in turn is converted into electrical energy using electricity generators.

The quantity of water available and its hydraulic head are two important parameters deciding the extent of electricity that could be generated from a given location. The quantity of water available at a site depends on the hydrological cycle of nature. The rain fall and the run off determines the site for the power plant. The maximum capaciy of the power plant is fixed based on the minimum quantity of water available.

2.1. Classification of Hydroelectric Power Plants according to Water Head Available

Based on the water head available, power plants are classified in the following manner :

- High head plants are those with a water head over 100 m. Turbines used are Francis/ Pelton wheels.
- Medium head plants are those with water head in the range of 30 and 100 m. Turbines used are Francis/Propeller/Kaplan.
- The indicative figure of water head available for a low head plants is 30 m. Turbines used are Francis/Propeller/Kaplan.
- Mini head plants are classified as the ones with head in the range of 5 to 20 m, and microheads are those with less than 5 m head.

It is to be noted that as we go down from high head to low head plants, we need to use more and more volume of water for generating the same quantum of electricity and accordingly, the turbines become bulky. There is also a danger that the very low head units belong to non viable category. This aspect need be given attention while we go in for mini/micro head plants.

It is considered worthwhile at this stage to refer to a report published in *The Hindu* of October 11, 2012 titled, "India to test run Japanese green projects". The report mentions that during our meeting with experts from Japan, we expressed interest on a mini-hydel project that generates electricity with a two metres drop of water. The reason for our interest on the same is reported to be due to the fact that the usual drop in our canals is around two metres.

This is a fit case for ensuring viability through detailed analysis before implementing the project.

2.2. *Energy intensity of hydraulic power*

Energy intensity depends on the water head available at the inlet to the turbine. This varies from one source to another. We have high, medium and low head sources. The locations are chosen ensuring that the energy intensity of stored water makes harnessing of electricity viable.

3. All Positives with Hydroelectricity

A well designed hydroelectric power station is the simplest, cheapest, environment friendly, as well as the most viable one. Major expenditure is incurred towards building the reservoirs and the power station. The start up cost for hydroelectric and thermoelectric power stations are similar at Rs. 50 to 70 million per MW of rated capacity. Smaller units have higher start up cost.

Unlike the case of thermoelectric as well as nuclear electric power plants, no fuel is used here. In addition, the life of dams and penstocks is over 100 years. These aspects enable most hydro-electric power plants to produce electricity at the cheapest rates.

On establishing viable units of hydroelectric power plants, we derive the following benefits:

- We shall be able to curtail the use of thermoelectric plants by a matching figure. Moreover, once established, these are eternal sources for the generation of electricity.

- This will also tell us the limit of hydroelectric potential in our country.

4. Hurdles in Setting up Power Plants

Decades ago, population density in mountainous regions was next to nil. Very few inhabitants were affected by the establishment of hydroelectric power plants, and they were relocated easily. Nowadays, fertile lands in mountainous regions have been largely occupied, especially in countries like ours.

Another issue that comes up while we propose to set up hydroelectric power stations is the concern raised by environmentalists. In our country, proposals for the setting up of hydroelectric power plants are met with resistance. This stands in the way of capacity addition to the extent of potential we are blessed with.

We established Bhakra Nangal around four decades ago. Probably its establishment had a smooth sail. We have recently established a power station in Narmada river. During its implementation, many issues were raised and some of these stood against smooth execution of the project.

At the same time, there are reports on the establishment of very large hydroelectric power stations in China, involving rehabilitation of more than a million. China implements such projects, since, they have the appropriate administrative machinery (ref: report titled "Displacement of natives of China for Hydro electric project", published in *The Indian Express* of January 22, 2010). The highlight of the report was that an additional 300,000 people will be forced to move out from their homes because of China's Three Gorges dam and its huge reservoirs. This was in addition to the 1.27 million people relocated earlier.

5. Installed Capacity

The installed capacity of hydroelectric plants in our country is around 35,000 MW (2010). The limit could be around 60,000 MW (assumption by this author)

6. Conclusions

- This is a green energy source that is eternal, and is in use for long.

- Power plants are available for procurement without restrictions.

- A well designed hydroelectric power plant produces the cheapest electricity. Once established, they serve us for a very long time. Dams and such other portions of the station last for over 100 years.

- Since hydroelectric potential is location specific, there are limitations in their capacity addition.

- Plants with high water heads are most viable for electricity generation. Creating installed capacities for electricity generation utilising this perennial and most viable source of energy to the maximum extent is an invaluable asset to any country.

- Establishment of new units is facing stiff resistance from environmentalists. This will adversely affect the scope for capacity additions we cherish to have.

- According to this author, a review in totality, considering all issues related to the establishment of electricty generating plants will provide full marks to hydroelectric plants, and also remove the hurdles we are facing currently for the establishment of new units.

Fission Based Nuclear Energy

1. Fission based nuclear energy is green

Significant quantum of heat/electricity could be generated from nuclear reactors. Since the quantum of hydrocarbon based fuels available is limited, harnessing energy from nuclear fuels is important and therefore, it might be worthwhile to use this green energy source for electricity generation judiciously.

India has been pursuing with the related studies for long. Even when some countries slowed down their work related to new areas, we did not backtrack.

Nowadays, we are seriously pursuing with augmenting power generation capacities from all possible green energy sources. The sources we have been successful in bringing into use are the power of flowing water, wind and nuclear fuels. Out of the three, hydro electricity generation has been the oldest. Wind electricity generation is catching up nowadays. Both these are location specific and there are limitations in capacity additions. Compared to these, nuclear power plants are not location specific, and they could be set up at chosen locations. This is one of its advantages.

2. Different forms of nuclear reactions

The forms of energy released from nuclear reactions are the following:

- In nuclear fission, when an unstable heavy nucleus is bombarded with high energy neutrons, it splits and large amounts of heat and kinetic energy are released.
- Fusion of light nuclei like H_2 and He into heavier ones is associated with the release of thermal energy. Sun is a fusion based nuclear reactor.

3. Fission based nuclear reactor

3.1. Functioning of a reactor

Uranium under normal circumstances, undergoes radioactive decay and in the process, emits neutrons and other subatomic particles. However, when uranium amounting to critical mass is kept together, chain reactions set in. This ensures continuous neutron emissions and release of thermal energy. Hence, in nuclear reactors, conditions for the initiation and maintenance of chain reactions is built in. Construction of the core is similar to that of a tubular heat exchanger, with fuel pellets assembled inside.

The heat generated is removed from the core of the reactor by circulating a heat extracting fluid such as water/heavy water/ liquid sodium around pellets. The heat of the steam generated is used to operate turbo electric generators.

3.2. Types of reactors

Nuclear reactors commonly used for power generation are:

(1) CANDU type reactors use natural uranium as fuel, and heavy water as coolant. Since no fuel enrichment is involved, this is suitable for countries without enrichment facility. However, since the power density is considerably low (9.7 kW/litre), the reactor size is large.

(2) Pressurised Water Reactor (PWR) is fuelled by enriched uranium and cooled by light water. Steam is generated in a steam boiler. It has high power density (65 kW/litre) and therefore compact.

(3) Boiling Water Reactor (BWR) uses enriched uranium as fuel and light water as coolant in a similar manner to PWRs, except for the variation that steam is generated in the reactor itself. The power density of the reactor is around half that is available from PWR. Both BWR and PWR were used by countries having fuel enrichment facilities.

4. Nuclear Fuels

4.1. Uranium

Some of the isotopes of uranium are the fuels for fission based nuclear reactors.

Uranium ore has around 0.714 per cent of U^{235}_{92} isotope which is fissionable and the non fisionable part is 99.286 per cent of U^{238}_{92} isotope. We value the fuel by the content of U^{235}_{92} in it. The more the percentage of U^{235}_{92} isotope in the fuel pellets charged in the reactor, we have higher thermal density inside the core.

4.2. Plutonium

During the functioning of a reactor, some atoms of U^{238}_{92} isotope are converted into plutonium (Pu^{239}_{94}). The fuel bundles removed from nuclear reactors are reprocessed for plutonium.

4.3. Thorium

Thorium (Th^{232}_{90}), exposed to fast neutrons in the reactor becomes a fissionable isotope of uranium, namely U^{233}_{92} and in this process, we derive energy as well as additional fuel simultaneously. The reactor in which this change takes place is called Fast Breeder Reactor (FBR). Since the quantum of fissionable isotope of uranium available as uranium deposits is limited, conversion of thorium into U^{233}_{92} isotope will provide additional fuel for powering nuclear reactors for a long time to come. Thus, the availability of fissionable materials is expected to increase manyfold by the large scale use of FBRs (see the brief on 'nuclear fuels' annexed).

4.4. Fuel enrichment

By adopting enrichment process, we increase the content of U^{235}_{92} isotope in the fuel bundles charged in nuclear reactors. Enrichment being a process that depends on the small difference in masses between the atoms of U^{235}_{92} and U^{238}_{92} isotopes, the efficacy of the individual enrichment stage is low.

The equipment used for one of the fuel enrichment processes is known as 'centrifuge'. Thousands of them are used in serial operating mode for achieving the desired percentage of U^{235}_{92} isotope in the fuel pellets. The process is a very elaborate one, and is highly energy intensive.

Media reports often indicate that serious objections are raised against the acquisition enrichment capability by new countries for the fear of proliferation of nuclear weapons. Enriched uranium has duel applications; either for fuelling nuclear reactors or in the atom bombs. This is the reason for the objections on the proliferation of enrichment plants.

5. Fast Breeder Reactor Development

Work on development of Fast Breeder Reactors is in progress. Achieving a viable breeding ratio is a challenge faced by those working in this area. Commercial units are expected to come into large scale use by around 2050 only.

India is pursuing serious research in this area. This aspect is evident from a report titled, "We won't compromise on three stage nuclear programme." published in *The Hindu* dated October 11, 2010. Following are the highlights:

- India has been actively pursuing a three stage nuclear programme. In the first stage, we used natural uranium and built 17 PWRs.

- In the second stage, plutonium reprocessed from spent fuel, depleted uranium and thorium kept in blanket form will be used as fuel to power a series of breeder reactors. Thorium is converted into U^{233}_{92} isotope.

- In the third stage, reactors use thorium and U^{233}_{92} isotope to generate electricity. Thus, the three stages are interlinked.

- The Fast Breeder Test Reactor (FBTR) which formed the corner stone for the second stage uses liquid sodium. The FBTR with 40 MW rating which has completed 25 years as of now, used plutonium – uranium carbide fuel. The reactor has paid rich dividends in the form of experience gained in design, fabrication and testing of advanced fast breeder reactor fuel.

- Beyond 2020, breeders of 1000 MW capacity are planned to be built.

However, it is evident that units on a commercial scale could be available only by around 2050.

6. Nuclear Electricity Generation Capacity (2010)

The Times of India dated 28th February 2012, carried a news item titled, "India may double its nuclear capacity by 2017". The relevant portion indicating the extent of electricity generated from fission based nuclear reactors is given below:

According to a study by the World Nuclear Association (WNA), 30 countries were generating nuclear power in the year 2010. The cumulative capacity of the 434 nuclear reactors globally is 2,630 billion KWh – which is 13.8 per cent of the global electricity generation. The table given below indicates the role of

Table

Country	*Quantum of electricity generated as percent of total*	*Country*	*Quantum of electricity generated as percent of total*
China	2.0	Brazil	3.0
India	3.0	United Kingdom	17.9
Russian Federation	18.0	United States	20.2
Germany	22.8	Japan	29.2
Finland	33.1	Korea	34.7
Chek Republic	35.8	Sweden	37.4
Slovania	38.2	Switzerland	39.2
Hungary	44.9	Belgium	51.7
Slovak Republik	54.7	France	75.1

Source: OECD, Department of Atomic Energy.

nuclear energy in meeting the electricity requirements of the countries referred.

The above information provides a broad idea on the contribution of nuclear reactors for electricity generation, both in India as well as abroad.

7. About Import of Clusters

We in India have been seriously pursuing with multidisciplinary research and development works encompassing a wide spectrum of activities belonging to nuclear electricity generation for many decades. We have indigenously built power plants of 200 to 500 MW, and have expertise in the related technologies.

It is believed that establishing nuclear power stations with clusters of imported units having a capacity of 1000 MW units could help us in achieving a quantum jump in electricity generation. Since we face shortage in the availability of uranium for fuelling the currently operated reactors itself, it is important that fuels in appropriate form are available for lifetime of the imported units from abroad.

These are the reasons for our requesting countries belonging to Nuclear Supplier's Group (NSG) for offering the necessary help to establish clusters of high capacity units. However, the related efforts met with stiff resistance from various quarters, and we even landed into a situation wherein, continuation of the then government was under threat.

The differences of opinions in pursuing with such a programme were primarily in the following areas:

* NSG wanted to oversee supply and operation of the units. Controls were related to the supply of fuels and their reprocessing aspects. Considering the fact that we in India have a stable government, and also since our past record of working in this area is clean and transparent, many among

us felt that some of the clauses related to the purchase contracts were discriminatory.

* Differences of opinions came up on fixing the responsibility for the malfunctioning of reactors if any, and compensating for damages.

* It was also pointed out that even on going in for installation of clusters as planned, they could meet only a minor portion of our electricity needs of the coming times.

The proposals for capacity augmentation as indicated above seems to have received a setback, due to the adverse climate existing internationally in the post Fukushima period.

8. Advantages of Fission Based Nuclear Power

Currently, thermoelectricity generating plants are largely fuelled by coal. In addition, derivatives of petroleum are used to fuel these. We are also aware that reserves of these fuels is coming down as we use them.

Nuclear electricity generation derives energy from nuclear fuels. The intelligent use of nuclear fuels is expected to provide a significant quantum of energy for a long time to come. In addition, this is a green energy source.

9. Limitations

We have many hurdles standing on our way of pursuing with the establishment and use of nuclear reactors for electricity generation. They include the following.

9.1. Set backs in nuclear power generation world wide

Accidents took place occasionally in nuclear power plants around the world. As a consequence, strong public opinion came up in many countries against their use.

Added to the same, we had an incident at Fukushima in Japan, wherein, a cluster of units landed into serious trouble. This major accident in which a large section of human population is reportedly affected directly or indirectly, has helped to add fuel to the fire. As an aftermath of the same, many are revisiting their decisions to go forward with nuclear electricity generation.

However, the recent pronouncements by Japan in this regard are confusing. They are reportedly working for a nuclear power free Japan by the year 2040. In the present times wherein todays decisions are changed by as early as tomorrow; when someone says that we will get rid of nuclear power 30 years later, one can not be blamed if he expresses doubt over such commitments and the motives behind such statements.

9.2. Nuclear reactor vs Atom bomb

For controlling fission reactions, nuclear reactors are provided with control rods made of neutron absorbing materials, namely, boron and cadmium. In the case of an atom bomb, when armed for explosion, the critical mass of the fuel is instantly created by bringing together individual masses, and uncontrolled chain reactions are initiated.

The fact that nuclear fuel in its enriched form is used both in nuclear reactors and atom bombs, creates disputes on the installation of nuclear reactors and fuel enrichment plants even for electricity generation. This stands on the way of utilisation of nuclear energy for peaceful purposes by countries who do not have the technological base for their creation.

9.3. Radio activity – a major hazard

Hazards due to radioactivity are given serious attention. This issue comes up during each and every phase of the activities involved and therefore, we generally ensure that elaborate safety features are built in. However, such provisions boost up the overall cost as well as make the equipments involved bulky.

However, in the light of the aftermath of tsunami in Japan, doubts are being raised on the capability of reactors to function in a safe manner in situations wherein, the units may have to face severe forms of natural calamities.

It would also be appropriate to recall a newspaper report wherein the difficulties we faced in our country on the development of a nuclear submarine were touched upon. They included the unwieldy nature of power reactors and the related inconveniences we had put up with during the integration of the nuclear power unit with the submarine. The bulkiness of nuclear power generation systems is the result of implementing design features for taking care of the mandatory radiation protection aspects.

9.4. Technologies Involved are Extensive and Complicated

The field of nuclear reactors is highly complicated with respect to their design, manufacture, operation and the like. Unlike the case of thermoelectric power plants, areas requiring R&D efforts are many. They include, mining and processing of uranium ores, enrichment of fuel and conversion into fuel bundles, operation of reactor, treatment of process wastes, long term storage/disposal of wastes having a radioactive life spanning to hundreds or even thousands of years and addressing to safety related issues involved from end to end.

The technologies involved in the above are complicated ones demanding continuous R&D in key areas involving large groups of scientists and technocrats. This is especially relevant in the current scenario, wherein, large commercial units are still under development in the laboratories and production facilities. In addition, the developments on FBR's are in their infant stages only as of now.

Many nations are not in a position to establish the infrastructure for carrying out R&D encompassing the large spectrum of science and technology related to the realisation of nuclear reactors. This may be seen in the light of the fact that units are not freely traded.

To be self sufficient in this area calls for heavy investments in the infrastructure for R&D laboratories, as well as production facilities for the manufacture of commercial units. This is a serious handicap with respect to the extensive use of nuclear reactors for peaceful purposes, particularly for countries not having the capability to produce such units.

9.5. Fuel Making is an Extensive and Expensive Process

Nuclear reactors are charged with fuel bundles once in a couple of year only. However, this does not make fuelling nuclear reactors a simple affair.

The reasons are:

- Few countries have uranium deposits in significant quantities.
- The fuel bundles are tailor made suiting to specific reactor designs. Therefore, in the case of imported units, the timely availability of fuel will depend on the success of fuel procurement contracts.
- Most fuel bundles involve enrichment; and the process is elaborate.

Ensuring that the fuel bundles are replaced in a timely manner is important with respect to ensuring high plant load factor.

It would be appropriate to highlight the fact that our country faced serious fuel shortages in the recent past (2010). Reports indicated that the situation improved subsequent to our signing of agreements with countries belonging to NSG and importing tailor made fuel bundles.

9.6. Safe storage/disposal of nuclear wastes

Safe storage of wastes with half life of thousands of years without causing environmental damages is a worrisome issue.

Sometimes it could so happen that nuclear reactors have ceased to exist on the face of the earth in a matter of a couple of centuries from now. In such a situation, there are no ways and means by which we could remove the burden of nuclear wastes from the responsibility of the coming generations. They might as well have to continue to ensure safe storage of the wastes, perhaps even without knowing its relevance. Even moral issues are involved here.

The aftermath of Fukushima accident of Japan in which many reactors landed into dangerous situation involving the letting out of radiations to populated areas around them, has resulted in changing the perceptions on the safe working of nuclear reactors.

10. Capacity Creation on a Long Term Perspective

Lack of adequate indigenous capability for producing reactors in large numbers and providing fuel as required in a timely manner stand on the way of accelerated capacity additions in our country.

Considering all the above, it is assumed in this study that we shall be able to achieve an installed capacity of 30,000 MW in a phased manner by the year 2050. It is also assumed that thereupon, we shall be able to continue to maintain this level of nuclear electricity generation thereafter.

In the light of concerns raised by the public on some of the proposals to establish nuclear parks in our country, it might be worthwhile to look into the issue from a different perspective. Perhaps, the same result could be achieved by establishing smaller units spread out widely in different parts of our country.

11. Decommissioning of Fukushima Nuclear Power Plant

The Hindu of December 23, 2011 carried a news item on the decommissioning of Fukushima nuclear power plant which was destroyed by tsunami on 11th March, 2011. Considering the relevance to our discussions, salient points are given below:

- The plant which functioned from 1970 was severely damaged due to the earthquake and tsunami. The decommissioning process is expected to last for 30 to 40 years.

- The damages are identified broadly as below:

  (1) Cooling system was damaged beyond redemption and caused the melting of fuel bundles charged in the reactors.

  (2) Radiation leaks occurred in a large scale.

  (3) Around 80,000 persons had to be relocated.

- The removal of the melted fuel is expected to go on for many decades.

- Many of the technologies associated with the dismantling operations need to be developed.

- There is less clarity on the expenditure associated with the dismantling process.

12. Conclusions

* Fission based nuclear energy is green.

* However, unlike the other green sources wherein the input power comes from the environment; nuclear reactors draw power from nuclear fuels, and therefore, the same is a non renewable energy source.

* The nuclear reactors currently in use are considered as viable electricity generating units.

* The viability of breeders which are under development as of now need to be acertained before using them on an industrial scale.

* Creation of fissionable nuclear fuel from thorium in the form of U^{233}_{92} in FBR's could enhance the availability of energy from fission based nuclear reactors.

* Currently, nuclear power reactors are used mainly for electricity generation and to a lesser extent, for district heating purposes in cold countries. These units are largely stationary. We could have used them widely in the transportation sector especially for propelling cargo ships, but for the safety related issues involved such as inadvertant radiation leaks and theft of nuclear fuels ending in their diversion for the making of weapons.

* The design, manufacture and operation of nuclear power reactors encompasses a wide sprectum of activities spread over many areas of science and technology and therefore, pursuing with the building of these units call for the effort of a large group of scientists and technocrats working in various disciplines. This also calls for the setting up of laboratories and manufacturing industries, and the associated development of know hows and know whys. These aspects are responsible for this area of activity being out of bounds to countries with scarcity of resources.

* Many of the solid, liquid and gaseous forms of materials used during the various stages of manufacture/operation of nuclear reactors emit radiations and hence, elaborate radiation protection activities are a part of any nuclear power programme. In addition, treatment/safe keeping of nuclear wastes becomes an additional responsibility bestowed on the scientists and technologists for many centuries later to the shutting of power generation units.

* For many of the developing/less developed countries, nuclear power is out of bounds due to the severe restrictions existing for the trading of related systems.

* Accidents took place in nuclear power plants in the past. As a consequence, there has been a slow down of research and development activities in many countries. This situation has been aggravated by the accident at Fukushima (2011).

Subsequently, uncertainties have arisen with respect to proposals for accelerated capacity additions in our country.

Wind Electricity

1. Wind energy is an eternal green energy source

In the past, non mechanised ships utilised for moving goods far and wide across continents, were powered by trade winds. This indicates that the power of wind was a valuable source of energy we made use of in ancient times.

Wind energy belongs to renewable energy category. Utilising this energy source to the extent of its potential will enable us to effect savings in the use of fuels.

2. Wind characteristics and wind electricity generation

2.1. Strong winds ensure viable electricity generation

We have a large wind electricity generating farm in the Southern Tamil Nadu. Thousands of units stand up majestically in this area. Most of these units have come up in a time span of a decade or so. The region has strong winds blowing across for considerable part of a year.

Majority of the world's wind turbines operate at places where the mean wind speed is around 7.5 $\pm$ 1 m/s. Most machines have a rated wind speed of 12 - 14 m/s. Rated wind speeds are generally considered to be in the range of 1.4 - 1.6 times the mean wind speed.

2.2. High level of fluctuations in wind speed

A recent report highlighted high level of fluctuations in the availability of electricity from wind farms. It also highlighted the difficulty experienced in planning for meeting the day to day demand for electricity due to extreme nature of unexpected fluctuations in the quantum of electricity generated by the farm. (Ref: report titled 'Wind leaves Electricity Board in lurch,' in the *Times of India* of July 8, 2010).

The report highlights the following with respect to functioning of the farm.

- The highly unstable supply of wind electricity has left Tamil Nadu Electricity Board (TNEB) officials confused. The drastic fluctuations in output ranging between 1000 MW and 3000 MW have forced the board to purchase additional power and maintain power supply.

- The peak period of wind electricity generation is between May and October every year. This year, until May 31, the average generation was 2000 MW, and in the first week of June it dropped down to 70 MW, but again touched 3000 MW.

- For the past two weeks wind energy has started fluctuating. On some days the generation is 1000 MW, and on some other days, 800 MW.

- Till June TNEB banked on wind energy and reduced power purchases. The drop in output is attributed to delay in the arrival of monsoon.

The related literature also indicates this difficult side of wind electricity generation. Being an eternal green energy source it is advantageous to use wind energy for electricity generation and hence, issues of this nature are addressed to by bringing into the grid and removing other types of power plants as called for.

2.3. Intensity of wind power and electrical output

The intensity of wind power is specified in watt/m^2. It is seen in literature that wind turbines installed at favourable locations convert wind power into electricity in the range of 400 watt per square metre of the blade sweep.

The power of wind varies in direct proportion to the cube of wind velocity. The following table illustrates the variation in the output power in relation to the wind velocity.

Table

Sl.No	*Wind velocity m/s*	*Cube of wind velocity m^3/s*	*Ratio*	*Output energy/ MW/year (plant load factor 0.25)*
1.	7.5	422	$7.5^3/7.5^3 = 1$	2.19 million kWh
2.	6.0	216	$7.5^3/6^3 = 1.95$	1.119 million kWh
3.	5.0	125	$7.5^3/5^3 = 3.37$	0.65 million kWh
4.	4.0	64	$7.5^3/4^3 = 6.69$	0.33 million kWh

Note:

(1) In the above table, the energy in the form of electricity output at a wind velocity of 7.5 m/s is compared with the quantity which is available for different wind velocities. At a plant load factor of 25 per cent, the output energy/MW/ year for a wind velocity of 7.5 m/s is 2.19 million units.

(2) It could be seen from the data given above, that the output energy in electrical units available from a wind electric unit at a wind velocity of 4 m/s is only around 15 per cent of energy output at a wind velocity of 7.5 m/s.

The above table indicates that when wind flows across the turbine blades, a volume of air passes over them. Therefore, cubic effect of wind velocity comes in the power of wind. This also tells us about substantial change in the energy output, even for a marginal change in the wind velocity.

2.4 Availability of wind for less duration and hence low plant load factor

We have discussed about the need for strong winds being available for installation of these units. Equally important is the duration for which wind with appropriate velocity is available. Any one who passes through a wind farm could see many units remaining in a sleep mode. It is natural to have doubts on the viability of these units for electricity generation.

However, it is worthwhile to note that no power plant runs all the time during its life. In the case of wind electric plants, when the wind speed is outside the range prescribed, the turbines are rested by the controls. For wind turbines installed at places where appropriate wind speed is available, the plant load factor is reported to be in the range of 25 to 30 per cent.

Unlike fossil fuelled power plants, we do not use fuel here. The savings in fuel cost compensates for the low level of plant load factor.

3. On Wind Turbine – Electric Units

There are horizontal axis and vertical axis turbines. The turbine blades are constructed to the shape of an aerofoil. The blades are usually twisted to maintain the angle of attack in the range of 4–8°. Tip speeds of blades are in the range of 50 to 80 m/s with peak efficiency for wind speed in the range of 7 to 9 m/s. Wind turbine design and manufacturing technologies have matured with the adoption of computers and tailor made softwares. A comparison among the earlier and the recent versions of wind turbines reveal these aspects clearly.

Most medium and large size wind turbines operate at constant speed, which is fixed by virtue of their being connected to fixed frequency electricity network, usually through a gear box and an induction generator. Direct drive wind turbines have no gear box, and use large diameter direct drive generators. They are provided with appropriate electrical systems.

The power at low wind speeds is insufficient to overcome loss of power in drive trains. Most machines do not rotate below a wind speed of 4 m/s. This is an important point with respect to site selection.

Presently, units are available in the range of a couple of MW rating also.

4. Start up Cost of Wind Electric Plants

According to the report titled "Contract on wind mills" published in the *Hindu Business Line* of Dec. 10, 2009, Rajasthan State Mines and Minerals Ltd. (RSMML) were awarded a contract amounting to Rs. 1,920 million for supplying 15 units of 2.1 MW each. This works out to Rs. 70 million per MW.

5. Installation of Units away from Shore Inside Sea

Wind electric units are sometimes installed a few kilometers away from the shore, inside sea. It is claimed that this sort of construction enables us to take advantage of the higher wind velocities prevalent in these locations.

As far as our country is concerned, locations with favourable characteristics for the installation of viable units are largely in hilly terrains. However, in cases wherein sea shore locations are suitable for setting up wind farms, it is sometimes considered advantageous to install the units offshore. In such cases multiple units are installed on floating platforms placed a few km inside the sea. Since offshore installations are costlier, the relative merits need be evaluated.

6. Some Island Nations Claim to have Extensive Wind Electric Potential

Reports highlight the advantage some island nations have with respect to wind electric potential. They claim that wind electricity could provide a significant portion of their electricity needs in the hydrocarbon depleted era.

If it is true, such nations could be considered as belonging to the category of the most blessed ones. This case is similar to the advantageous position of countries blessed with significant hydroelectric potential.

7. Subsidies for Promotion of Generation Capacity

In order to promote accelerated capacity additions in the area of wind electricity generation, governments have been providing subsidies over the procurement price, waiver of taxes and duties, as well as helping by entering into liberal power purchase agreements. However, subsidies provided for the promotion of wind electricity generation is small compared to the level of subsidies provided for solar electricity generation. Being an important green energy source, this promotionary aid provided could bring down the cost of the units, as well as bring in improvements in the related technologies.

8. Wind Electricity Generation Potential of Our Country

In the absence of appropriate data, the wind electric potential in our country is assumed to be in the range of 50,000 to 60,000 MW for the purpose of our discussion. However, taking into consideration the difference in plant load factors and working out the equivalence of this installed capacity to that of hydro/thermal/nuclear power plants, this works out to 21,000 to 25,000 MW.

9. The Recent Media Reports on Wind Electricity Generation

(1) About units installed at Wagamon, Kerala

Subsequent to the installation of a wind farm with an installed capacity of 33 MW at Wagamon situated in the Western Ghats region, the State government faced many issues related to the manner in which the land wherein the farm was built. The issues were largely related to the ownership of land which belonged to Adivasis and the manner in which rights for installation were obtained.

(2) About units installed in the State of Maharashtra

The Hindu of February 29, 2012 carried a news item on the issues reported by Western Ghats Ecology Panel, related to the installation of wind electric units in the State of Maharashtra.

The highlights are:

(a) The wind farm installed near Bhimshankar Wildlife Sanctuary has an installed capacity of 113 MW, and involved an expenditure of Rs. 7,720 million.

(b) The 194.66 hectare area of the farm covers 14 villages, and is situated in the reserve forest.

(c) This area is the habitat of Malabar giant squirrel which is the State animal of Maharashtra.

(d) The creation of the wind farm has resulted in severe environmental damages.

Note:

(1) The installed capacity of the above mentioned wind farm is only equivalent to hydro/thermal/nuclear power production units of $\frac{113 \text{ x } 25\%}{60\%} = 48$ MW. When we mention about the installed capacity of these units, we do normally indicate this aspect.

(2) A forest land of 194 hectare was used for installation of the units which has a hydro equivalent generation capacity of around 48 MW only. The Athirapally hydroelectric project of Kerala (163 MW) which is withheld from execution, covers only around 150 hectare of forest land. Hydroelectric projects belong to premium class green electricity generation projects (according to this author's evaluation), however, it seems that we do not have a system that views the related issues in totality and provides green signal on such aspects.

(3) On wind electric units installed in the State of Tamil Nadu.

A report on the wind farms of Tamil Nadu appeared in the *Times of India* of June 2, 2012. The salient points are given below.

* The installed capacity as of now is 7034 MW.

* Due to the onset of monsoon, the output is in the range of 2600 - 3500 MW. It is envisaged that we shall receive significant output from these until September.

* The units are installed at Muppanthal of Kanniyakumari district, Panagudi of Thirunelveli, Kayathar of Thuthukudi and Palladam of Western Tamil Nadu.

* Out of the demand for around 12,000 MW in Tamil Nadu, the current level of distribution is only around 8,000 MW. The fact that wind farms could provide electricity to the extent as indicated above atleast during the monsoon season tells us clearly about the importance of harnessing the power of wind, by installing units wherein they could function in a viable manner.

10. Conclusions

* Installation of wind electric farms aimed at harnessing the power of wind is a recent phenomenon. This is an eternal green energy source.

* Technologies of wind electric units have matured, units are traded without restrictions.

* Capacity additions are subsidised by the government at different stages for promoting its wide spread use, as well as for enabling technology upgradations.

* Capacity additions are limited by the wind electricity potential one has.

* Wind electricity generation units could be established only in regions with strong winds.

* A small variation in the wind velocity results in large variations in quantum of electricity generated by the units. This highlights the need to ensure the availability of wind at the site with minimum velocity while we go forward with establishment of the units.

* The limit of wind electric potential of our country is assumed to be in the range of 50,000 – 60,000 MW for deciding upon our green energy potential. Modifying this figure by taking into consideration the low plant load factors to which these units work (25 per cent), this installed capacity is only equivalent to 21,000 – 25,000 MW installed capacity of hydro/thermal/nuclear electricity generating plants.

* Before we install units inside sea away from shore, we need to evaluate the advantages compared to installing them onshore.

* Media reports indicate that some island nations have large wind electric potential to the extent that the same could meet a significant portion of their electricity needs. In case this is true, such nations belong to the category of blessed ones, since, in the hydrocarbon depleted era they could meet the energy needs of their economy largely by utilising eternal green energy sources.

* Many units are installed in mountainous regions resulting in serious environmental damages.

4

New Green Energy Sources

Studies on harnessing electricity from new green energy sources are being carried out in many parts of the world. They include energy from nuclear fusion, solar electricity, geothermal electricity and electricity generated from ocean waves/tides.

Deriving energy from fusion based nuclear reactors is a very complicated matter.

Solar electricity generation is used for isolated applications for a long time. However the recent trend in our country is to promote capacity additions by establishing grid connected solar electricity generating plants. At present, many governments are promoting capacity additions in this area by sharing the financial burden in the form of subsidies for the power plant equipments, waiver of taxes and duties, loans at reduced interest rates, and providing heavily subsidised power purchase agreements.

Utilising geothermal energy as well as the energy of ocean waves/tides are limited by their potential for building viable electricity generating units. The related issues are discussed in the following chapters.

Energy/Electricity from Nuclear Fusion

1. The need for deriving energy from nuclear fusion

Fusion based nuclear reactors are considered to be one of the main sources of heat/electricity for the future. Once we are able to demonstrate commercial units working in a viable manner, we shall be able to exploit this green energy source to a substantial extent.

2. What is nuclear fusion?

Nuclear fusion is a process by which light elements such as hydrogen are fused together into heavier elements. As a byproduct of the fusion reaction, thermal energy is released. Brief on various combinations of fusion reactions is given in the annexure (see the write up titled "Nuclear fuels" annexed).

However, nuclear fusion is a process that is extremely difficult to be demonstrated even on a laboratory scale.

Sun is believed to be creating energy by the fusion of hydrogen nuclei into helium. However, unlike the simple processes we use for deriving energy from different energy sources, nuclear fusion calls for creating an environment in which a pressure of billions of atmospheres and temperature in the range of 100 million °C exists. The nuclei of isotopes of hydrogen are expected to fuse together by overcoming the nuclear forces of repulsion between the protons of the fusing atoms. The enabling force for the fusion to take place is the thermal energy present in its core, and the process is therefore known as thermo nuclear fusion. As far as the sun is concerned, these unbelievable conditions have come into existence naturally. The unimaginably high level of gravitational pulling force that exists in the sun is believed to be enabling the fusion process to take place in a container less situation.

The controlled and sustained nuclear fusion reaction is achieved by heating the hydrogen isotopes and creating a plasma that then must be confined to a doughnut shaped device called 'tokamak'.

3. Fuel for fusion based nuclear reactors

Most fusion reactions fuse hydrogen isotopes, namely, deuterium and tritium. Deuterium is found in sea water to the extent of 30 mg per litre. The production of deuterium from sea water involves the separation from it of heavy water (D_2O) - or water where hydrogen atoms of normal water (H_2O) are replaced by deuterium. The heavy water is then subjected to electrolysis to yield deuterium. Tritium is produced in nuclear reactors by the action of neutrons on lithium.

According to the above mentioned information, fuel for fusion powered reactors available on earth are eternal, however, viability aspects of fuel processing need be understood. At the same time, as per the details of experiments to be conducted during the Moon missions planned by space faring nations, an isotope of helium (He_3) is proposed to be mined, processed and brought to earth for fuelling fusion based reactors we plan to create.

The reason for earmarking the isotope of helium as a preferred fuel for fuelling fusion based nuclear reactors could be the following.

It is true that fusing hydrogen nuclei into helium is what is being fundamentally talked about as the fusion reaction meant for generating electricity in a large scale. In nuclear fusion reactions, the kinetic energy of the nuclei undergoing fusion is to overcome the electrostatic repulsive forces of protons in the same. The order of magnitude of temperature required to overcome the repulsive forces between protons is 10^8 °C. Different combinations of nuclei are proposed for the nuclear fusion. Hydrogen - hydrogen is the least preferred one. Deuterium - deuterium reaction comes next. The most preferred one is tritium - deuterium, this is said to take place at a comparatively lower temperature (Both deuterium and tritium are isotopes of hydrogen). May be, the use of helium isotope planned to be brought from the Moon could fuse easier; similar to that between tritium and deuterium.

It is possible that conclusions could be drawn on such matters only once we make some headway on the fusion related experiments we are pursuing with currently.

4. Facilities Built for Nuclear Fusion Related Experiments

Information available on the experimental units set ups by different agencies are given below.

4.1. International Thermonuclear Experimental Reactor (ITER)

The description of the environment in which the fusion reactions take place has provided some insight in to the difficulties we face in formulating fusion related experiments on a laboratory scale. We have reports regarding nuclear fusion related experiment which is formulated as part of International Thermonuclear Experimental Reactor. This is considered to be one of the world's largest collaborative R&D projects. A fully superconducting 'tokamak', ITER is designed to produce about 55 MW of power from a plasma that is sustained for 400 - 500 seconds.

China, India, South Korea, United States, Soviet Union, European Union, Japan are signatories to ITER.

4.1.1. Experimental set up

The relevant aspects of a report published in the recent past is quoted here. The report provides brief information on an experimental setup planned on nuclear fusion.

The title of the report is, "A star is born; first step to fusion power".

A glimpse on the working principle of the experiment is as follows. A giant array of lasers will allow harness the process that powers sun in controlled and sustainable miniature nuclear explosions.

Inertial Confinement Fusion (ICF) uses multiple lasers to rapidly compress and ignite isotopes of hydrogen.

Features of main equipments and accessories are as below:

- *Target chamber*: This is a vacuum chamber of diameter 10 m. Its wall is of composite construction. The inner shell is made of 100 mm thick aluminum, and the outer shell is of 300 mm thick concrete. Weight of the aluminum shell is around 80 tonne, and the concrete shell is 240 tonne. Boron in casing absorbs X-rays and neutrons.

- *Fuel*: 150 microgram of frozen tritium and deuterium is enclosed in plastic, carbon or beryllium sphere and kept in a thermal control package that maintains its temperature between 18 K and 20 K.

- *Lasers*: 192 laser beams fire 1.8 million joules burst sufficient to light five billion 100 W bulbs for 20 billionth of a second. The power of lasers put together works out to 10 kW.

Lasers converted into Ultra Violet light inside of gold cylinder absorbs U.V. energy and radiates X-rays.

4.1.2. Functioning of the experimental setup

- Lasers fire as low energy pulses from oscillator, amplified billions of times and split into 192 intense laser beams.

- X–rays bathe and rapidly heat the surface of the fuel capsule.

- Surface expands, compressing the core to 100 billion atmospheres.

- Fuel core density exceeds 1000 g/cm^3. This is 100 times the density of lead.

- Thermonuclear ignition of fuel at 100 million °C, hotter than the centre of sun.

Energy is released by fusion of tritium and deuterium atoms. Mini star lasts for fraction of a second, but should emit 10 to 100 times energy used to ignite it. Only byproduct is harmless helium.

The above description could give a fair idea on the realisability or otherwise of a workable reactor.

4.1.3. First firing of lasers

Highlights of report published in *The Times of India* of November 18, 2010 on the then status of the experiment are given below:

- A team at the National Ignition Facility, Livermore, California has fired up the 192 laser beams at the centre of the reactor and aimed them at the glass target containing tritium and deuterium gas.

- The resulting release of energy was of a magnitude of 1.3 million mega joules, and the peak radiation temperature was approximately 6 million °F, say the scientists. For a direct comparison, the temperature at the centre of the sun is 27 million °F.

- Scientists estimate that the 'proto fusion reactor' could be operated by the year 2020, and by the year 2050, almost quarter of the U.S. energy supply could be produced from this energy source.

The above report indicates the status of the very preliminary level of experiment undertaken by us in this area.

4.2. Chinese experimental setup

This is from a report in the *Hindu* of June 8, 2007, authored by Pallavi Aiyar.

According to this report, China's Experimental Advanced Superconducting Tokamak (EAST) is the world's first fully superconducting Tokamak that can potentially sustain the generated plasma for 1000 seconds, compared to the tens of seconds most other tokamaks can manage. For fusion to become commercially viable in the future, it is necessary to develop methods to sustain plasma for far longer.

EAST which took just five years to build, is a smaller version of ITER. China hopes that this unit will enable ITER fusion researchers to get a head start on learning to tame plasma for longer periods of time. The first demonstration of plasma lasted for nine seconds.

4.3. India and South Korea

Also according to the above report, India's Steady State Superconducting Tokamak (SST – 1) is a smaller version of EAST. However, South Korea's Superconducting Tokamak Reactor (KSTAR) will rely on superconductors made from alloys that are even more advanced than those used in EAST.

5. A Different View on the Realisation of Fusion Based Power Reactors

The report by Pallavi Aiyar also indicates that generation of energy out of fusion based nuclear reactors could simply be a pipe dream, according to some of those who are familiar with the intricacies of creating a fusion based power reactor. It compares this effort with the success we had in the field of fission based nuclear reactors, wherein, the time period between the discovery of fission, and it being put into use as a power source was just three years. It points out that fusion is still to make that transition more than half a century since its discovery. The report quotes Chinese experts and points out that it could be another 30 – 50 years before the world can expect fusion to begin to generate commercial results.

6. Analysis

The two different assessments from those who manage the related developmental efforts as quoted here are the following:

* A statement made after the first firing of lasers of ITER, wherein, those associated with the experiment claimed that by the year 2050 a quarter of the U.S. energy requirements could be met from fusion based nuclear reactors.

* Another is from an expert who manages the Chinese experiment who has opined that it would be 30 - 50 years from now before the world can expect fusion to begin to generate commercial results.

(According to this author, perhaps both these conclusions are made without going into the intricacies involved in the creation of a fusion based nuclear reactor unit.)

Keeping aside the experiments we have undertaken, it seems difficult to find answers to the following common sense questions on deriving energy from a fusion based nuclear reactor.

- The first one is on the creation of a condition with high temperature and pressure required for the fusion. In the core of our sun we have the reaction sustained in a containerless situation. Creating such an environment in a container of the types we are familiar with on earth is beyond imagination. Even the virtual containers we create (this is what is done for the experimentations undertaken until now), essentially have to be made of down to earth materials and systems we are familiar. How to create a system in which continuous nuclear fusion reaction takes place (at temperatures in the range of 8 - 40 million °C and billions of atmosphere pressure) without causing damage to the very systems that create and maintain the environment for the fusion reaction to take place, looks to be an extremely difficult question to answer. It is rather easy to say that the temperatures and pressures required for a nuclear fusion reaction could be created and maintained by using the available techniques; however, achieving the same is another. Nature has such a creation in the form of our sun. The energy is flowing out of the sun at a rate that is mind boggling. It might not be an easy task to find a solution to this seemingly unsolvable problem.

- The processes involved in deriving energy out of such a system lack clarity. The methods we use to derive electricity from the various categories of power plants we use are simple and are well developed. Water, steam and steam turbines are common factors in the thermal as well as nuclear power plants we use for the generation of electricity. Their operations involve at the most pressure of the order of a few hundred atmospheres, as well as few thousand degrees Centigrade of temperature.

7. Conclusions

* Solar energy is produced as a result of fusion of light elements such as hydrogen into heavy elements.

* The gravitational pulling force existing on sun is believed to be around 100 times the force existing on the surface of earth. This is believed to be the reason that light elements such as hydrogen are kept together in the core of sun enabling fusion reactions to take place incessantly.

* Our developmental efforts on nuclear fusion reactors could be summed up as containerizing a 'container less reaction' existing in nature.

* Since our example for a functioning fusion reactor is sun itself, some of us seem to believe that with the unbelievable and unimaginable levels of advances we believe to have achieved in science and technology, we shall be able to create mini suns on earth in the form of fusion based nuclear reactors and derive energy out of them.

* However, the description of an experimental set up of a very preliminary nature points out to the extremely high level of complications involved in the creation and maintenance of workable units. It is to be noted that these experimental

setups under creation in various laboratories are not complete reactors, but only intended to demonstrate the basics of nuclear fusion.

* Considering the above, it would be wise to include this energy source as a contributor to the future energy scenario only after a viable commercial unit has been demonstrated.

Solar Electricity Generation

1. Getting to Know About Sun and Earth

Sun is our prime source of energy and is responsible for the evolution of life. We shall briefly mention some of the statistics of the Sun and Earth.

Table

Feature	*Sun*	*Earth*
Mass	1.991×10^{30} kg	5.97×10^{24} kg.
Radius	696,000 km	6,378 km
Average density	1,400 kg/m^3	5,520 kg/m^3
Surface temperature	5,762°C	-

Average distance of earth from sun is 150×10^6 km.

Sun radiates energy all around and a very small portion of the same falls on earth. The surprising fact is that in spite of the continuous incidence of solar energy for billions of years, the temperature on the surface of the earth has remained in the range that enabled the evolution and sustenance of life. It is evident that earth sends back into space nearly all the energy it receives from the sun.

2. Sun is a Fusion Reactor

The process by which sun produces energy is interesting to learn. The solar interior constitutes the main mass of the sun, and has gases at pressures of billions of atmosphere, and temperature in the range of 8 x 10^6 °C to 40 x 10^6 °C. The density in this region is in the range of 80 to 100 times that of water. The sun is a fusion reactor and the constituent gases are being retained by gravity.

We could also see nature's beauty in such a phenomena. Many manifestations of nature are beyond the comprehension of even those of us who claim to be scientifically and technologically advanced twenty first century humans.

Sun's energy is available to us in the form of electromagnetic radiations. Our atmosphere screens out a portion and lets the balance to earth's surface.

3. Availability of Solar Energy

3.1. Solar constant

We express the power of sun's radiations in terms of kilo watt/m^2 or kilo joule/second/m^2. Here we have 'solar constant'; the energy (joule) received per unit time (second) on a unit surface area (m^2) kept perpendicular to solar radiation in space just outside the atmosphere. Its value is 1353 W/m^2.

3.2. Solar energy available on earth

Atmosphere reduces the intensity of solar radiations to around 1000 W/m^2; this value is for tropics, on a clear sunny day around noon.

Table: Insolation at select locations in India

Location	*Latitude*	*Mean horizontal insolation kWh/ m^2/day*	*Yearly insolation kWh/m^2*
New Delhi	28 degree 35′ N	5.40	1973
Kolkata	22 degree 39′ N	4.84	1767
Jodhpur	26 degree 18′ N	5.94	2169
Pune	18 degree 32′ N	5.59	2065
Chennai	13 degree N	5.59	2040

Many factors reduce the quantum of solar radiation available on the surface of the earth to 1500 to 2000 kWh/m^2/year. Data on the yearly solar insolation at some locations of our country is given in the following table.

4. Conversion of Solar Energy into Electricity

Methods for the generation of solar electricity are Solar Photo Voltaic (SPV) and solar thermal. First, we shall look into the various aspects of Solar Photo Voltaic method.

4.1. Solar Photo Voltaic (SPV) Systems

4.1.1. Solar Photo Voltaic electricity generation

When sun's rays fall on solar cells, electricity is generated without the involvement of mechanical generators. Here is an example of direct conversion of solar energy into electricity. This is due to 'photo electric effect' which is defined as the generation of electromotive force as a result of absorbing radiation. Energy conversion device used to convert sunlight into electricity using photoelectric effect is called solar cell.

A photo voltaic system consists of the following:

- *Solar array*: Consists of solar cells grouped together and made into flat panels; converts solar energy into D.C. electricity. Normally these panels are fixed to ground.

- *Blocking diode*: Lets the array generated power to flow only towards battery or grid, as programmed.

- *Battery*: Stores when the generated electricity is not used. Two types of batteries in use are lead acid and nickel cadmium. Lead acid batteries used in cars and trucks are not the appropriate ones, however, these are the ones used nowadays as a part of solar electricity generating systems.

- *Inverter*: Changes the DC output to grid compatible AC of suitable voltage, frequency and phase.

- *Power conditioner*: Since the voltage output of the PV module varies with the quantum of solar energy incident on it, this varies from time to time. Systems with battery storage require voltage or shunt regulator to prevent overcharging of storage battery.

- Appropriate switches and circuit breakers are part of the system.

The optimum operating voltage of a photo voltaic cell is generally 0.45 V. For an exposed area of 40 x 10^{-4} m^2, the current would be around 1.08 A. Accordingly, the output power = 0.45 V x 1.08 A = 0.49 W, in full sunlight. The cells are connected in series and parallel, thereby deriving power of specified characteristics.

4.1.2. The extent of solar electricity available at the utility point from SPV systems

Some of the factors responsible for reduction in solar electric power output by the PV systems are given below.

* We have the solar constant whose value is 1.353 kW/m^2. The atmospheric transmission factor brings it down while the solar rays cross earth's atmosphere and reach earth. For m = 1.0, the factor is 0.633, and the same comes down to 0.276 for m = 5.0. The transmission factor depends on the length of rays passing through the atmosphere. Therefore, for a factor of 1, we have the intensity of solar radiation on the surface as 1.353 x 0.633 = 0.825 kW/m^2.

* Then we have the conversion efficiency of the PV module. For various reasons, around 45 per cent of the energy of solar radiation is converted into electric current in the chip. Due to factors such as resistance of the semiconductor material and some energy being reflected back into space, the actual efficiencies of conversion into electricity by the PV modules are much lower. The best single crystal could be considered

to yield an efficiency of around 15 per cent only. So we have come down from 0.825 kW/m^2 to 15 per cent of the value i.e. 0.825 x 0.15 = 0.124 kW/m^2. (124 W/m^2)

Thus, at this stage of the discussions, the energy available as electricity by photo voltaic conversion has come down from the attractive figure of the intensity of solar radiation outside earth's atmosphere which is 1353 W/m^2 to 124 W/m^2.

* Then we have another major factor that brings down the energy available further. We are looking for the electricity available at the utility points, it could be a compact fluorescent lamp, a ceiling fan, or a cooking oven.

Earlier we had listed down the parts of a solar photo voltaic system, namely, storage battery, inverter, blocking diodes, switches and circuit breakers. All these have their individual efficiencies. On compounding, if we get 60 per cent of energy which is converted by the PV module at the utility points, its value could be around 124 W/m^2 x 0.60 = 76 W/m^2

4.2. Solar Thermal Electricity Generation Systems

4.2.1. Electricity generation

In solar thermal method, sun's rays are concentrated and directed on water circulated through pipes. On absorption of solar heat, water changes into steam. This in turn is used to drive steam turbine - electricity generating units. According to a report that appeared in *The Hindu* of May 23, 2010 titled, "Alstom makes foray into solar energy," the solar thermal system proposed for supply by the company has a central receiver for receiving sun's concentrated heat energy with the help of thousands of mirrors focused on it and thereby steam is generated. This steam in turn drives a steam turbine - electricity generating unit.

4.2.2. Viability related issues on solar thermal electricity generation

The effort involved in the setting up of the solar thermal power plants could be no different from that of solar photo voltaic power plants. This is evident, since, we have to derive sun's heat by using thousands of mirrors.

We can understand more about viability of solar thermal method from the following real life examples.

4.2.2.1. Heat intensity available from LPG Stove

15 kg of LPG fuel burned over 100 hours by a LPG burner and consuming 0.15 kg/hour of gas, transmits 6,270 kilo joule (1.74 kW) of thermal energy to cooking vessel. Now, if the cooking vessel's bottom is of 0.25 m dia and area is 0.049 m^2, the intensity of heat available for cooking is

$$= \frac{1.74 \text{ kW}}{0.049 \text{ m}^2} = 35.5 \text{ kW/m}^2$$

The above example indicates that under the best possible conditions, the intensity of direct solar energy is only around 3 per cent of the thermal energy intensity we get from the LPG stove referred above.

4.2.2.2. Comparison of intensities of solar and wind energies

The wind turbine electricity generating units of the present day, convert wind energy to the extent of 400 W/m^2 of the turbine blade sweep into electricity. This happens at around a mean wind velocity of 7.5 m/s, and this conversion factor has a major role in wind electric units becoming viable. The figure of 76 W/m^2 we have arrived at for the direct solar energy harnessing is a small percentage of this. This is one factor which could contribute to non viability of solar electricity generation by the photo voltaic route.

5. Some Factors Limiting the Efficiency of Solar Photo Voltaic Modules

The factors include:

- Efficiencies of PV modules or arrays are lower than that of solar cells, since, areas of a solar module in between the cells are left without utilisation. The fraction of the areas of cells to the area of the module is called the packing factor. This is a factor responsible for bringing down the efficiency of solar panels.

- The production of a free electron – hole pair in a semi conductor requires a specific amount of energy. In silicon for example, the minimum energy is 1.1 eV. This energy is not available in radiation having a wavelength of more than 1.1 micro meters. Thus, the infrared radiation having longer wave lengths is unable to produce photo voltaic effect in silicon material. Energy from this radiation is largely absorbed as heat. Even in the case of radiations of other wave lengths, the energy in excess of that is required for the production of an electron – hole pair comes out as heat. In the case of silicon, only around 45 per cent of the energy of solar radiation is capable of producing electrons and holes.

Other factors include the following:

- Losses due to the electrical resistance of the semi conductor material.
- Part of the solar energy is reflected back to the sky.
- Some energy is absorbed by the non photo voltaic surfaces.
- In the case of modules and arrays, there are losses due to mismatch between the individual cells in a module as well as between the modules in an array.

6. Price of Solar Electric Plants

* The price of plants with an installed capacity of 3 MW rating established in the State of Karnataka in 2009 was around Rs. 200 million/MW.

* L&T established solar electric plants at their premises during the year 2009. The price of plants with an installed capacity of 406.0 kW rating was around Rs. 100 million. This works

out to around Rs. 250 million/MW (ref: report in *The Hindu Business Line* dated November 15, 2009).

* Mar Athanasious College of Advanced Studies, Thiruvalla, Kerala has installed a solar electric plant rated at 30.0 kW, at a price of Rs. 7.2 million. (ref: report in *The Hindu* of November 25, 2011). They received Rs. 2.2 million as subsidy from the Ministry, and Rs. 3.7 million as low interest loan refinanced by National Bank for Agricultural and Rural Development). The output of the plant was expected to be 45,000 units per year. This works out to Rs. 240 million/ MW.

- Scheme for electricity generation and distribution using 'Roof top solar electric plant' was the title of a news item published in *The Malayala Manorama* daily of March 9, 2011. According to this report, the price of solar electric plants proposed for installation was around Rs. 2.0 lakh a kilo watt. This works out to Rs. 200 million/MW.

- A catalogue distributed on March 2012 at Thiruvananthapuram, in connection with the sale of solar electric units as a part of Jawaharlal Nehru National Solar Mission indicated the price of a 200 watt, 24 volt unit as Rs. 75,000. This works out to Rs. 375 million/ MW.

- *The Hindu* of October 4, 2012 carried a news item titled 'Poojapura jail goes solar'. The report was on the eve of switch on of a plant meant for solar electrification of the Central Jail, Thiruvananthapuram. The plant is designed to cater to the requirements of street lighting, lighting and fans for blocks, steam cooking, chappathy making unit, and water pumping. The report highlighted that the plant with an installed capacity of 229 kW is set up at a cost of Rs. 79 million. (The above reported expenditure of Rs. 79 million for an installed capacity of 229 kW works out to Rs. 345 million/MW. Perhaps this high level of expenditure is due to the fact that the electricity generated by the solar plant is upgraded so that the same supports a variety of electrical drives with high rating. This is evident from the description of the equipments powered by the plant).

- *Malayala Manorama* Daily of April 10, 2013 carried a news item titled,' 100 Marks for Vimalagiri Public School'. The report highlights the following with respect to a solar electric plant installed at Vimalagiri public school, Moovattupuzha, Kerala:

- The school has installed a 37 kW solar electric plant at a cost of Rs. 100 lakhs. This works out as Rs. 270 million/MW.
- The solar panels numbering 159 are reported to be having a total area of 5,000 sq.ft.
- Batteries numbering 144, weigh of 4,000 kg.

It could be seen from the data spanning a period of four years from 2009 to 2013 that the price for solar electric plants has not come down during this time.

7. Our Unrealistic Expectations Regarding Solar Electricity Generation

Many books are available on utilisation of solar energy. In addition we often see reports on the same in media. All of them say in one voice that solar energy is the energy of tomorrow. What we infer from this statement could be that in the times when we finish utilizing the hydrocarbon based fuels such as coal and petroleum, we are sure to derive as much energy as required in various useful forms from solar energy and support our then existing economy. We need to evaluate this aspect with all the seriousness it deserves.

We utilised solar energy in the past; we shall be utilising it in future too. However, it seems that we have taken our expectations regarding the utility of solar energy sky high, particularly our expectations on solar electricity generation. With this sort of expectations, we seem to have landed into a dangerous situation. Our assessment on the viability related issues pertaining to the production of solar electricity for grid connected applications indicates that we need to have a realistic evaluation of the related issues and reframe our expectations regarding the same.

There was a recent media report which referred to a government's proposal to make installation of solar electricity generation systems mandatory on residential buildings of more than 3,000 square feet built up area. From the energy saving point of view, it would have been wiser to ban the construction of such large residential buildings for single family.

8. Conclusions

- Intensity of solar rays as available outside earth's atmosphere is around 1353 watt/m^2. Regions above equator receive solar energy totalling to 1500 to 2000 kWh per sq.m.

- Solar panels available for commercial use provide around 76 watt electricity per sq. m at utility points (this figure is worked out by the author). Compared to the quantum of electricity provided by wind electricity generating turbines per sq. m of blade sweep which is said to be around 400 watt/m^2, the quantum of energy converted by the solar panels into electricity is substantially low.

- In the past, solar electricity was being utilized for specific and non grid applications such as solar panels for satellites and providing electricity at remote places. Many factors which remained unfavourable with respect to solar electricity generation kept us away from utilising it for on grid applications.

- "Jawaharlal Nehru National Solar Mission" is a programme of the Government of India adopted a couple of years ago. The programme envisages that the installed capacity of solar electric power plants in our country is raised to around 20,000 MW by 2022. The units could belong to both grid connected as well as non grid connected categories.

- Since the economic factors involved in setting up and operating these plants were not favourable, our government decided to provide heavy subsidies at all stages. Providing support for solar electricity generation as above is aimed at bringing down the cost of solar electricity generation to the level of expenditure incurred on setting up plants such as thermoelectric ones in the long run. However, many related factors are seen to be against achieving the above mentioned goal. In addition, since the plant load factor of solar electricity generating units is around 15 per cent only, a 20000 MW installed capacity is equivalent only to around 5,000 MW installed capacity of elelctricity generating units belonging to the categories of hydro/thermal/nuclear energy sources (solar electric plants are able to function only for around four hour out of every 24 hour; this is the main reason for the plant load factor being only around 1/4th of hydro/thermal/nuclear electric power plants).

- Life of solar photo voltaic plants is the lowest among the electricity generating plants. It is claimed that panels have a life of around two decades. At the same time, the life of storage battery, which is one of its expensive sub system, is only in the range of 3 – 4 years. Media reports of the times during which the plan came into being mentioned that depreciating the capital spent for installing solar electricity generating plants in a short time span of seven years was agreed upon.

- In comparison to the heat intensity we get from fuels of biological origin, the intensity of solar rays falling on solar panels is very low. Further, the efficiency of conversion into electricity is also low. These factors together are responsible for very low output from solar electric power plants.

- For upgrading the viability of electricity generating units, nowadays, thermo electric/nuclear electric plants are made to highest rated capacities. Units in the range of 1000 MW rating are the order of the day. Even in the case of wind electric units, the power of single unit has come up to a couple of MW. In comparison, we are unable to make solar electric units compact. A one MW rated unit necessarily is to have around 10,000 m^2 of exposed area of solar cells, that too in units numbering around 500 to 1000, with all related accessories for the individual units. This plays a major role in makir g the solar PV power plants extensive.

Geothermal Energy

1. Geothermal Energy Source is Green

Geothermal energy is a vast and eternal energy source available in the earth's crust. It is believed that this can be utilized in the form of thermal/electrical energy in a big way. It is also believed that we need to put in serious efforts towards harnessing electricity from this source in a large scale.

2. Characteristics of Geothermal Energy

Earth's interior is believed to be at around 4000°C, and has iron and nickel in liquid form. The temperature gradually comes down as we move towards its surface. Its outer surface is rocky, with an average thickness of around 80 km. The temperature of rock increases as we go down from the surface at a rate of around 25°C per km. This indicates that the temperature at a reasonable drilling depth of around 4 km is too low to serve as a heat source for geothermal pow er plants.

3. Why Exploit Geothermal Energy Source?

The utilisation of geothermal energy is being advocated nowadays in the context of our growing concern on the adverse effects of burning petroleum and coal in large quantities. Efforts have been put in by few countries with advantageous geothermal features for its utilisation by establishing pilot plants.

4. Locations for Extracting Geothermal Heat in a Viable Manner

We have locations on the earth's crust with substantially lesser thickness compared to the average and therefore higher thermal gradients across the crust. Also, at some of these regions the molten magma breaks through to the surface of the earth as volcanic eruptions. The above locations are favourable for harnessing geothermal heat.

Warm thermal springs resulting from the natural heating of ground water occur at some places. We also have locations where the near surface rock has steam trapped in huge quantities which could enable industrial scale utilisation. This steam could be used for the heating of buildings in cold regions. In regions where appropriate quality steam is available, the same could also be used for the generation of electricity.

5. Extracting Geothermal Energy from Hot Rock

For creating the geothermal heat extraction facility and generating electricity from steam, we provide the following:

- Wells are drilled to a depth of few km in the rock and large volumes of rock are fractured at its bottom. Heat extraction fluid which is usually water is passed through the bores, and geothermal heat from the fractured portion of the rock is extracted. Hydraulic fracturing, fracturing by subjecting to thermal stressing, chemical and nuclear explosive fracturing are experimented.
- Steam – electric power plant established at the borehead converts the thermal energy of steam into electricity. In

some facilities, the steam extracted from the geothermal well is converted into superheated steam, using external heating facility.

However, the creation of geothermal wells with efficient heat transfer characteristics by fracturing the rock underneath the bore, seems to be a task with uncertainties. In addition, the chances of adopting nuclear fracturing techniques is an impossible task; especially in countries like ours wherein a lot of issues of uncomfortable nature are raised by our international friends on explosion of any sort of nuclear device.

6. Geothermal Source and Environmental Conditioning

Since many of the European countries are in the northern hemisphere beyond tropics, for most part of a year the temperature of the environment is substantially less compared to our body temperature. Many of these countries have what is known as the 'cool summer' and they are required to use for most part of a year significant quantum of fuels for heating up their homes, factories and public utility buildings. Use of even low quality steam tapped underground is helpful in bringing down the demand for fuels in countries belonging to these regions.

7. Viability Aspects of Geothermal Units

Viability of geothermal units need be ensured before going in for them in a large scale.

The reasons are:

- Steam is available from geothermal wells at low temperatures and pressures. The efficiency of a steam electric power plant is high only when high quality steam is used. While we get an overall efficiency of 30 – 35 per cent for steam – electric power plants fuelled by fuels, a geothermal heat generated steam – electric power plant generally could have overall efficiency in single digits only. This low efficiency brings down the quantum of electrical units output from the power plant.

- A significant portion of electricity generated by the plant need be used for the operation of equipments for pumping water for heat extraction.

- The start up cost per MW of rated capacity could be significantly higher, compared to power plants set up with conventional energy sources.

- The size of a single geothermal power plant is reported to be in the range of few hundreds of kW only. This low capacity of individual units could make them less cost effective. Generally units of steam - electric power plants fuelled by coal/naphtha are in the range of hundreds of MW.

8. No Viable Sources in Our Country

It seems that we do not have locations with favourable geothermal features in our country which could provide high quality steam for running electricity generating power plants.

9. Conclusions

- In general, geothermal heat is a weak energy source.

- Some countries with favourable geothermal potentials are reported to be harnessing energy in the form of heat, and to a lesser extent as electricity.

- Countries with cool summer are the appropriate places for tapping geothermal heat, since, it could be used directly for the heating of buildings.

- In our country, locations wherein geothermal energy could be used for generation of electricity in a viable manner seem to be non existent.

- No data seem to be available on the viability of geothermal – electric units established until now.

Electricity Generation from Ocean Waves/Tides

1. Ocean Wave/tide Energy is Green

It is generally believed that the energy contained in ocean waves/tides could be advantageously used for generation of electricity. This belongs to eternal green energy category.

2. We have a very long coast line; but..

Our country has a coast line stretching to more than seven thousand kilometers.

However, locations with favourable characteristics essential for harnessing electricity from ocean waves/tides need be ensured by undertaking viability analysis.

3. Experimental Unit near Thiruvananthapuram

Some countries have established experimental stations and generated electricity out of them. We in India had also put up an experimental station at Vizhinjam near Thiruvananthapuram and carried out investigations.

A report published at the time of winding up of the above mentioned unit is referred here (Ref: report titled "Harnessing electricity from ocean waves" published in the *Malayala Manorama* daily of November 30, 2009).

Highlights of the report are:

- The experimental station was the first of its kind in our country, and we were the fourth nation to go for the same.
- The developmental effort was planned and led by IIT, Chennai.
- The electrical power capacity of the station was 150 kW, and could burn 7,500 compact fluorescent lamps rated at 20 W.
- The unit was commissioned in the year 1991, and the same was used to power a desalination plant in the year 2000.

- The main part of the unit was a concrete structure known as *'caison'*. Salient features of the structure are:

  (a) It was as tall as a five storied building, and weighed around 3,000 tonne. Its three sides were closed.

  (b) The side that was open faced the waves.

  (c) Waves entered the 'caison' and energised the electricity generating device.

  (d) Caison is reported to be of a short life of 20 years only, and therefore the authorities dismantled the structure in the year 2009.

The cost incurred for building the unit rated at 150 kW is reported to be well above Rs. 3.0 Crore (at 1990 price level).

It is sure that those associated with the experimentation have got a volume of data from this experiment. It seems that the unit was not put under commercial production giving an indication that it was not a viable one.

The requirement of such a massive structure for tapping the wave energy indicates the following:

- In the process of deriving energy, very large volume of water contained in the waves/ tides has to be handled.

- Trade off studies on materials of construction of caison would have narrowed down on concrete. Alloys might have been excluded due to prohibitively high cost, or corrosion related issues getting prominence.

- Since the caison had to handle water from a large area, the concrete structure is required to be stiff and therefore heavy.

- The intensity of potential/kinetic energy contained in the waves is very low.

The start up cost of the experimental unit comes to around Rs 200 million per MW of rated capacity (at the price level of 1990's). This is very high for an electricity generating power plant. This could be a factor that decides its viability.

4. Unit at Portugal

A report regarding a power plant established at a sea coast of Portugal was published in *The Hindu* of September 6, 2008 under the title, "Waves of clean power from Portugal". We shall have a look at this unit.

At the heart of the station built inside the sea 5 kilometers away from the coast are three cylindrical wave energy converters. Moving up and down on the endless wave of the open sea they convert its power into elecricity. At the peak output, the wave machines will generate 2.25 MW, enough for the annual needs of 1500 homes.

Each of the semi submerged cylindrical tube is 142 m long, and has a diameter of 3.5 m, and is made of 700 tonnes of carbon steel. A single wave converter consists of four articulated sections that pass up and down, as waves pass along it. At each of the hinges between the sections, hydraulic rams use the wave motions to drive electricity generators to produce upto 750 kW of power at the peak output. The electricity generated by the three devices is carried by undersea cables to a sub station, which will then be fed to the national grid. This is reported to be the first commercial type of power station established at Portugal.

The economic aspects of this set up is not available.

5. A feel on its economic viability

In the absence of data on the economics of this unit, a feel on its viability could be got from the following:

Since we have often visited sea shores, we know that the waves have the following characteristics:

- The intensity of energy contained in them is very low.

- Waves are of intermittent nature. Between momentary kicks available from the rising waves, there is a wide gap. This characteristic results in very low plant load factor for the electricity generating unit.

- Because of the low energy intensity, large volumes of water need be handled for enabling power production.

- Rating of 2.25 MW alone does not provide us with an understanding on the quantum of ouput energy available from the unit in terms of kWh.

The above characteristics boost up the cost of the power plants, and in all probability, make them unviable as electricity generating units.

6. Conclusions

* Harnessing electricity from ocean waves/tides has serious limitations.

* Energy of ocean waves/tides is intermittent in nature, moreover, its intensity is low. Therefore, very large volume of water is handled for deriving a specified quantum of energy. In addition, its plant load factor could be very low. These characteristics make the electricity generating plant bulky, unwieldy as well as costly.

* The viability of individual units harnessing electricity from this source need be evaluated and concluded.

* Locations with favourable characteristics have to be chosen for the establishment of viable units. This seriously limits the potential of this energy source.

* We could bank on ocean wave/tide electricity generating projects for the future, only after their viability is proven. Therefore, as of now, we are bound to exclude this energy source while planning for our future energy scenario.

5

More Solutions to Mitigate Climate Change

In addition to the role of green energy sources in the future energy scenario about which we have discussed so far, certain other related developments are being pursued in the area of harnessing green energy in the desired forms and their utilisation.

These are:

- Adoption of hydrogen economy in which hydrogen is used as fuel.
- Production of bio-fuels in a large scale and use for fuelling various transportation systems.
- The wide spread use of solar cooker is being discussed with the intention of bringing in savings in the use of hydrocarbons.

Each one of these is discussed in the following chapters.

Adopt Hydrogen Economy

In hydrogen economy, the energy for motive power or generation of electricity is derived from hydrogen. Serious developmental efforts are undertaken in laboratories related to various aspects pertaining to its realisation, with the goal of gradually switching over to hydrogen economy.

Presently, internal combustion engines fuelled by petroleum distillates are the prime movers in mobiles we use on land, water and air. They include engines of light duty and heavy duty vehicles used for road transport, diesel railway traction, ships and aircrafts.

In addition, extensive electricity distribution network powers a large variety of industrial as well as domestic equipments/machines. Major portion of electricity is generated by burning coal. When we switch over to hydrogen economy, a new energy conversion device known as fuel cell functioning on hydrogen as fuel is proposed to be used in place of the internal combustion engines. In addition, we intend to make these as capable of working on hydrogen as fuel.

1. Need for Changeover to Hydrogen Economy

The major reasons are:

- Burning fuels for the past couple of centuries has increased the level of carbon dioxide (a greenhouse gas) in our atmosphere. This is believed to allow less of solar energy to return back into space and thus result in warmer environment on earth.

- This warming of our environment on a global level is believed to be already taking place and is said to be evident from the melting away of Arctic ice at higher rates compared to the past. In addition, the onset of unpredictable weather at higher frequencies is also feared.

- Arctic Ocean is a huge reservoir of fresh water in the form of ice; melting of which could increase the level of oceans around the globe to the extent that many metroes across continents could be submerged.

- Even a marginal increase in temperature of the environment is expected to have adverse impact on the yield from crops.

Serious debates are taking place in international forums on cutting down the emission of carbon dioxide. This virtually tells us that we have to burn less and less of fuels such as coal and petroleum in the coming decades. As the economy of developing countries have to keep growing at reasonably high rates for many more decades, this reduction in the consumption of fuels can be implemented only by compensating for the short fall from green energy sources. In this context, changeover to hydrogen economy is considered as beneficial.

3. Hydrogen Economy and Global Warming

The product of combustion of hydrogen and oxygen is water. By switching over to hydrogen as fuel, we might get rid of production of carbon dioxide during the energy conversion processes. This will have marked improvements on the quality of the environment around us. The utilisation of hydrogen as fuel is proposed to be implemented by using tailor made IC engines as well as fuel cells.

4. IC Engines Fuelled by Hydrogen

IC engines with some modifications, are said to function satisfactorily with hydrogen as fuel and are reported to be under evaluation. The main hurdle faced with respect to utilisation of hydrogen in IC engines is related to its storage. Since hydrogen exists in gaseous form it could be stored in mobiles by two different ways.

They are:

- Compress hydrogen gas to a pressure of hundreds of atmospheres and store in a fuel tank at room temperature.
- Liquify hydrogen and store in cryogenic tanks.

Adopting either of the processes results in very heavy hydrogen storage tanks. This is one of the factors standing on our way of using hydrogen in mobiles.

5. Fuel Cells

5.1. Working principle

A fuel cell is a device in which controlled electrochemical oxidation of fuel produces electricity. A fuel cell has anode and cathode separated by an electrolyte which is acidic or basic. Chemical reaction between fuel and oxidant takes place at the electrodes and they also act as conductors for electron flow in the external circuit. Migration of hydroxyl ions from cathode to anode enables the transfer of electrical charge through the electrolyte within the cell. Continuous operation requires the removal of heat, water and inert materials which enter the cell with the reactants. Platinum on the electrode surfaces act as catalyst for the electrochemical reaction.

Electricity available from a fuel cell is D.C. For ensuring grid compatibility, this electricity need be converted into A.C.

Therefore, a fuel cell system has the following sub systems.

- A fuel processor that converts the available fuels into hydrogen.
- The fuel cell.
- An inverter that converts D.C. into A.C.

5.2. Severe working conditions of fuel cells

Fuel cells function in an environment involving high temperatures and pressures. This makes their designs at least one order more complicated compared to that of the traction batteries currently in use.

5.3. Concern – inadequate life of fuel cells

Life of any system that we use has a direct relationship to its economic viability. Let us look at a familiar example. While we use petrol/diesel cars, nowadays, we also promote ones powered by traction batteries. In the case of petrol/diesel powered cars, the engines support them till the end of their life. At the same time, the life of a traction battery is 3 – 4 years at best, since, the same depends on the number of times the same is charged/discharged. The case of fuel cells is no different either. It is already pointed out here that fuel cells are one order higher in complication and therefore, ensuring their satisfactory life is also a challenging task.

5.4. Concern over conforming to net energy analysis criteria

It seems that the fuel cells, according to their current status of development do not pass the 'net energy analysis criteria'. For meeting the above said net energy analysis criteria, the quantum of energy required for its manufacture and life time operation need be only a small part of the quantum of electricity available from the same during its life time operation. Fuel cells have to meet this criteria; then only they will be able to bargain for the place of internal combustion engines in mobiles.

6. Hydrogen Production

6.1. Sources

We have most hydrogen existing on earth in chemical combination with oxygen in the form of water. In addition, all biological substances have hydrogen in chemical combination with carbon and oxygen.

6.2. Production

Hydrogen could be produced by adopting the following processes:

- *Steam refining or partial oxidation of coal*. The process is widely known as coal gasification.
- *Produce from liquid petroleum sources*.
- *Electrolysis of water*: Electricity provides the input energy for the process. In this case solar electricity is proposed to be used.

6.2.1 Production from coal and liquid petroleum, and concern over high cost

Leaving aside the environmental issues related to the production of hydrogen from hydrocarbon based fuels such as coal, tar sands and shale oil, there are no uncertainties on the use of the processes. However, the prevailing high cost of hydrogen indicates the possibility of the production processes becoming non-viable. Presently, high cost is acceptable, since, hydrogen is used in chemical industries for the manufacture of products belonging to essential category.

6.2.2. Production of hydrogen by electrolysis of water

Electrolysis of water involves its splitting into constituents namely, hydrogen and oxygen and is enabled by passing electricity through the electrochemical apparatus. The proposal is to electrolyse water using solar energy and use the hydrogen thus produced in fuel cells. The above proposal seems to be highly non-viable for the following reasons:

- We have already discussed and concluded that solar electricity generation is a non-viable process.
- We use solar electricity for electrolysis, and the hydrogen produced through electrolysis is used in fuel cells. This combination of processes is again a non viable one.

7. Carbon Sequestration as Solution to Limit Environmental Pollution

In order that the environmental damages on global level are addressed, it is proposed that the carbon dioxide produced during the manufacture of hydrogen is captured in bulk and stored/disposed elsewhere in such a manner that the same does not get into atmosphere. The related process is known as carbon sequestration.

The potential storage sites include spent oil fields, saline aquifers, layered basalts and deep sea ocean. The deep sea injection of carbon dioxide is proposed to be carried out at a depth greater than 3500 m, this is the depth at which carbon dioxide becomes denser than sea water. The carbon dioxide injected reacts rapidly with sea water to form solid clathrate, which is denser than liquid carbon dioxide and sea water. Hence, the injected carbon dioxide could end up in sea water as slush. This would gradually dissolve, releasing carbon dioxide into the surrounding sea water where it would react with dissolved carbonate and borate ions to become chemically bound in the form of bicarbonate ion. As the concentration of carbonate and bicarbonate ions is small, neutralisation would take place gradually as the carbon dioxide rich sea water mixes with the surroundings.

Carbon sequestration process is in its proposal stage only, and therefore, various aspects including viability of using the process on commercial scale need be proven.

8. Conclusions

* We desire to switch over to hydrogen economy. This is aimed at change over to clean energy forms such as hydrogen and electricity. Both these are considered to be the preferred energy carriers of the future.

* We also look for flexibility in the use of energy conversion devices: Fuel cells used for powering mobiles could also be used to provide electricity for home while they idle in garages.

* Prototypes of fuel cells as well as IC Engines working on hydrogen as fuel are reported to be undergoing performance evaluations.

* Hydrogen in gaseous form could be used in the internal combustion engines, and for the same, modifications need to be carried out in the related systems.

* As of now, fuel cells do not pass 'net energy analysis' criteria.

* Being similar to batteries in construction, fuel cells have limited life compared to internal combustion engines. This

has its adverse effect on the economics of this energy conversion device.

* Economics of hydrogen economy are not yet proven.

* Hydrogen does not exist in free form in nature. Therefore, this fuel need be manufactured in large scale primarily from hydrocarbons for enabling the adoption of hydrogen economy, and while we venture into the same, there is a need to make the hydrogen production processes environment friendly.

* Since hydrogen has to be manufactured using chemical conversion processes, cost of fuel thus produced is very much on higher side as of now. It is necessary that this is brought down to significantly lower levels for enabling the adoption of hydrogen economy. However, the possibility of cost reduction is a question mark as of now.

Production of Bio Fuels

1. Bio Fuel is a Green Energy Source

Plants utilise carbon dioxide from atmosphere and grow and while we burn their parts, the same is released back into atmosphere. Thus, bio fuels belong to green energy category. The present day energy managers seem to believe that bio fuels could be produced and used in a large scale in the coming times.

2. Brazil and Bio Fuel Production

2.1 Favourable aspects for bio fuel production

Points relevant to Brazil on bio fuel production are given below:

- Brazil produces sugar and ethanol in large volumes. With respect to volume of sugar output they are the second in the world, behind India. Favourable factors include vast areas of land suitable for sugar cane cultivation.

- Henry Ford, the pioneer of automobiles predicted in the year 1925, that ethanol manufactured from sugar cane could be fuelling automobiles in future.

- Compared to India, Brazil has 2 1/2 times land area and at the same time, their population is only 1/6th of ours. This aspect enables them to dedicate vast stretches of arable land for cultivation for bio fuels.

2.2. World War - 2 was used as an opportunity to augment bio fuel production in Brazil

During World War - 2, countries faced shortages for petroleum. Many went in for coal liquifaction plants and manufactured liquid fuels in a large scale, and thereby managed to keep their war machinery oiled. However, Brazil went into producing bio fuels from sugar cane. They blended up to 65 per cent of ethanol with petroleum based fuels. Later on, their ethanol manufacturing industry had its ups and downs. However, decades ago, Brazilian government stepped in and framed policies that ensured continuous growth in ethanol production.

Today, around 50 per cent of Brazil's economic output is reported to be from sugar cane/ethanol production.

2.3. Brazil makes flex cars

Currently, Brazil makes flex cars, they work on petrol and ethanol mixed at varying proportions.

2.4. Contract with the U.S.A. for exporting ethanol

On realising the need to enhance the use of bio fuels, the U.S.A. is reported to have entered into a contract with Brazil for large scale ethanol production and supply. The quantum of ethanol Brazil agreed to supply is many times their current level of domestic use.

2.5. Indications of Brazil largely becoming green in the coming times

This also tells us another aspect of Brazil's energy self sufficiency. During our discussions on hydroelectricity and wind electricity production, we found that there are countries with potential to the extent of meeting a major portion of their energy needs through these renewable and green energy sources. It seems that Brazil also comes under this category. Their hydroelectric, wind electric and bio fuel production potentials put together might meet a significant portion of their energy needs in the era of petroleum/ coal depletion.

The above aspect indicates the conditions which would enable a country to maintain their economic progress in times of hydrocarbon depleted era.

3. Need for Foresight in Integrated Energy – Population Policy

At the same time this also tells us that the absence of integrated energy – population policies will have its toll in countries like that of ours in future.

This is particularly so, since, no debates are taking place on the longevity of liquid petroleum deposits and the measures that need be taken up on its depletion. The present perception looks to be the blind belief that in the petroleum depleted era, we will have a smooth transition to fuels/energy from other sources, they include solar energy, hydrogen economy, bio fuels and the like.

However, this does not look to be based on evaluation of viability of the energy conversion processes which are perceived to be promising for the future.

4. Seeds as Sources for Bio Fuels

We discussed sugar cane as the source for bio fuels. However, it is perceived by some that bio fuels can be produced in large scale from castor, soyabeen seeds.

5. Viability Aspects of Bio Fuel Production through Net Energy Analysis

5.1. Need for assessment

Fuel/energy production processes have to be invariably assessed with respect to viability aspects, this is applicable to bio fuel production also. Since bio fuel production activities demand energy to be used at various stages, assessment of energy balance is absolutely essential.

We have debated in later part of the book on the net energy available from various types of electricity generating power plants. The outcome of the study includes the following:

- The net energy available from different types of power plants vary. This indicates the viability aspects of each of these.
- We do not get net energy from all types of electricity generating power plants and therefore, we should not persume that all of them to provide useful service to us. An example for the same is solar electric plant.

On the same lines, the availability of excess energy or the lack of it from bio fuels also need assessment.

5.2. Article by Sidhartha Varadarajan addresses viability aspects

In this context, it is considered appropriate to quote from an article published in *The Hindu* of June 2, 2007; this is authored by Sidhartha Varadarajan. The article provides the following information relevant to our discussions:

- According to the study conducted by UNICA on the bio fuel production from sugar cane by Brazil, the ratio between the energy spent for cultivation and production of bio fuel put together and the energy available from the fuel thus produced, is of the order of 1 : 8.9. This advantageous situation is also due to the fact that parts of the sugar cane plant are dried and used as fuel, for enabling the various processes involved in the bio fuel processing.
- At the same time, the report says that the same ratio with respect to soyabeen or castor seed is in the range of 1:1.3 to 1.8 only. This indicates the non-viability of bio fuel production from these.

- The bio fuel production is encouraged by governments by providing subsidies. Large scale diversion of inputs such as fertiliser, electricity and other services to bio fuel cultivation is seen to make them dearer, which in turn makes the life of those pursuing with cultivation of food crops miserable.

6. Conclusions

* Production of bio fuels from sugar cane is considered to be viable. This source could be exploited in a big way by countries having suitable excess land in matching proportions.

* India is in the forefront with respect to sugarcane cultivation. However, unlike Brazil, we do not have the availability of vast stretches of unused land which could be used for sugar cane cultivation. More over, the continuing high rise in population of ours is sure to demand higher quantum of production of food in future. This is expected to strain even the presently used land for the same.

* According to studies carried out on viability aspects, sugar cane is the promising source for the same. Sources such as castor and soyabeans are not found to yield excess energy to a satisfactory level, so as to make the bio fuel production from them viable.

* Since the excess energy available varies among types of plants, it becomes essential to evaluate each category with respect to viability aspects.

Use Solar Cooker

1. A Device that Utilises Green Energy

The Department of New and Renewable Sources of Energy is advocating the use of solar cooker in our country. Government wants to encourage the use of solar cooker, since, its wide spread use can result in fuel savings. It is good to probe into this matter.

2. A Cooker for Domestic Use

Some aspects of a solar cooker which is apt for meeting the cooking needs of a small family are given here. It is of double walled box construction made of wood or metal, and of size 0.6 m x 0.6 m x 0.2 m. The top has two glass sheets placed one over the other with spacing in between. Inside bottom of the box is painted black, and made leak tight at the top. A vent is provided for vapour to escape during cooking. A plain reflective mirror to the size of the box is hinged at the top. This is to increase the collector area and in turn provide higher temperatures for cooking.

Temperatures inside the cooker is in the range of 70 to 110°C above ambient, with a maximum of 140°C in winter and 160°C in summer. Metallic vessels used for cooking are provided with tight lids and painted with dull black on their outer surfaces.

3. Cooking Time for Dishes

Rice, dal and vegetables take 0.50 to 2.0 hour, and meat takes 2 to 4 hours for cooking. Cooking is faster between 10 am and 2 pm, when the solar intensity is high.

4. Limitations in Use

4.1. Cooking possible only around noon

Since temperatures needed for cooking are available only for few hours around noon, cooking has to be done during this segment of the day only.

4.2. Dishes that need frying is out of bounds for solar cooker

Solar cooker is not fit to prepare items such as poori, chappathy, dosa, vada and the like. This is, since, the intense heat required for cooking these is not generated in a solar cooker.

4.3. Need for planning

On careful planning, one can cook dinner on clear sunny days. As far as lunch is concerned, in case we are cooking a vegetarian one, it might be possible. For the finishing touches on curies, we have to use a stove energised by fuels. One need to also plan dishes in advance and they have to be bunched together for cooking.

Another possibility for using solar cooker is that we leave the items to be cooked for dinner before leaving for work. The cooked items could be collected in the evening, carry out finishing touches and serve. This could provide some leisure time in the evenings, as well as savings in fuel.

5. Conclusion

Stress is being given globally for substantial reduction in the consumption of fuels and extensive use of green energy. Here in comes the need to utilise solar cooker. Any portion of cooking carried out by solar cooker will result in fuel savings. In case we modify our food habits and use of solar cooker to the extent feasible, the same would result in conservation of firewood, LPG, Kerosene and coal as the case may be. Solar cooker could take care of a portion of our cooking needs in the coming centuries.

6

Coal Liquefaction and Gasification

1. Why Go in for Coal Liquefaction/Gasification?

Nowadays, deliberations take place regarding the adoption of coal liquefaction/ gasification processes in a large scale. The reasons for these are given below.

Petroleum is a non renewable energy source and therefore, as we use, its reserves deplete. Considering the rate at which we have been depleting petroleum reserves during the past one century or so, points out to the unpalatable reality that its availability might drastically come down in the near future. We assume that we will face this dangerous situation by the turn of the twenty-first century itself. Currently, a significant portion of our mobiles are powered by petroleum distillates in liquid form, and to a lesser extent in gaseous form. Once their availability comes down, we are sure to face serious difficulties, particularly in the transportation area.

Coal is believed to outlast petroleum. Initial phases of the development of automobiles saw coal fired steam engines powering railway cars. Gradually, we switched over to petroleum fuelled transportation modes. Further down, railway cars are upgraded to electricity propelled ones. However, liquid/gaseous forms of petroleum continue to fuel mobiles we use on road, sea and air.

Therefore, in the petroleum depleted era, it becomes imperative that we convert coal into liquid and gaseous fuels and feed our economy. This is also essential, since, the use of coal in solid form in the road transportation sector is sure to cause atmospheric pollution to unacceptable levels. Then we have the aircrafts, it is essential to have liquid fuels to power them.

A statement made in the year 2008 by the Chairman of British Petroleum indicates that crude oil might be available only for around four more decades. Even if it outlasts this period, it might not be wrong to assume that we could be facing serious shortages in liquid fuels by the end of twenty-first century. We make this assumption here and go ahead with our discussions.

Along with petroleum products coal has been meeting a major share of our energy needs. Nowadays coal based thermoelectricity generation is the primary source for electricity. The carbon dioxide produced during the combustion of coal as well as petroleum products is believed to have a major role in causing warming of the environment in which we live. Its adverse effects are believed to be setting in on us gradually and threatening our very existence.

Both the aspects mentioned above, namely, possible depletion of petroleum as well as the perceived threat of global warming are prompting us to look for the manufacture of synthetic fuels from coal.

2. Hydrogen Production through Coal Gasification

Hydrogen is considered as an ideal fuel, and could be produced in a large scale by coal gasification process. It can be used to fuel almost any device. The product of combustion of hydrogen is water and therefore, hydrogen is environment friendly. In addition, calorific value of hydrogen is the highest among all known fuels. All these positive aspects have prompted us to look at hydrogen as a fuel for the future. Because of this background we are making serious efforts for switching over to hydrogen economy. Since the liquid and gaseous forms of petroleum are being consumed by us at high rates, coal is an important source for hydrogen production.

Coal gasification processes involve chemical reaction of coal at high temperature with steam/air/oxygen to produce a mixture of gases consisting of carbon monoxide, carbon dioxide, hydrogen, methane, and hydrogen sulfide. Hydrogen sulfide is an undesirable gas produced from reactions with the sulfur present in coal. By reacting with water, carbon monoxide gets converted into hydrogen and carbon dioxide. Carbon dioxide and hydrogen sulfide are removed in the purification system. The reaction of hydrogen with carbon monoxide could provide methane gas. Thus, we get hydrogen and methane from these processes.

3. Processes are Over 200 Year Old

In the early nineteenth century, water gas and town gas produced from coal were used for illumination. Germans produced liquid fuels in a large scale from coal during World War II.

4. Processes are Not Environmental Friendly

We have already seen earlier that carbon dioxide is a byproduct of coal liquefaction/gasification processes. Scientists are working to evolve methods of capturing and disposing the carbon dioxide which comes out during coal liquefaction/gasification processes. One such proposal is to store it in empty spaces created by large scale mining of coal. Another method is known as carbon sequestration, a method of disposal at ocean beds. Both these disposal methods could be expensive, their viability as well as implement ability on a large scale are yet to be proven.

In addition, there are other environment related issues which include:

- Water in the form of steam is a raw material to coal liquefaction/gasification processes. A large coal liquefaction/

gasification unit will need enormous quantity of water for its functioning and could deplete ground water levels.

- Both air and water could be polluted by coal dust.

It seems that associated environmental issues are serious in nature to the extent that they could stand in the way of our adopting the coal liquefaction/gasification processes in a large scale.

5. Chemical Processes of Coal Liquefaction/Gasification are Elaborate

Petroleum refining is carried out by utilising the differences in their volatility, and the process is known as fractional distillation. Because of the low level of pressures and temperatures involved in the process, refineries are simple in their design and operation. Feeding the crude as well as removal of the distillates at different stages do not present engineering related difficulties.

On the other hand, coal liquefaction/gasification are chemical processes. The pressures and temperatures involved are high by any standards. The processes of feeding raw materials and collecting products are technically challenging and demand high level of engineering design inputs. Materials used for the construction of process equipments have to withstand the high pressures and temperatures and hence, they are made of special materials and alloys. All these could have direct bearing on the cost of capital involved.

However, in a situation wherein the liquid and gaseous petroleum reserves have been depleted by us, we have no choice other than deriving these from coal and meeting the fuel requirements of transportation industry.

6. Impact of Non-availability of Liquid Petroleum on Air Transportation

The issue of non-availability of petroleum distillates for air transportation is worth debating. For debating on the issue we will have to first decide as to whether there are possibilities of any new type of propulsion systems coming in for powering aircrafts. We are aware that new types of devices are being proposed for powering cars and buses in the form of fuel cells. However, no such proposal is being talked about with respect to air transportatio... Therefore, we could presume that suitable fuels in liquid form have to be provided as replacement for aviation kerosene from non-petroleum sources.

There looks to be two options for liquid fuels in liquid petroleum depleted era, namely:

- Produce liquid fuels from coal.
- Produce bio fuels.

Once the availability of aviation kerosene has drastically come down, we might adopt a combination of the above and operate our air transportation systems at a scale that is controlled by the quantum of liquid fuels available then.

Issues involved in the production of bio fuels in a large scale have been discussed in another chapter. We have concluded that the availability of bio fuels on a large scale could be difficult in the light of the higher level of human population that need be supported by the farming sector in the coming times. This is particularly applicable to countries like ours. Even today we find it difficult to find additional cultivable lands which would enable us to have a take off with respect to our dream called, "Second green revolution".

7. Conclusions

* It is generally believed that coal will outlast petroleum. Therefore liquid/gaseous fuels produced by adopting coal liquefaction/gasification processes could meet the demand for liquid fuels in the petroleum depleted era.

* Adoption of coal liquefaction/gasification processes in a large scale enables us to manufacture hydrogen and methane. It is proposed to create systems functioning with hydrogen as

fuel; this is intended to reduce atmospheric pollution caused by the burning of fuels.

* However, developmental efforts aimed at fructifying fuel cells as well as hydrogen economy in which hydrogen is proposed to be used as fuel are in their infant stages only as of now. In addition, these efforts seem to be meeting many road blocks.

* Unlike the processes by which petroleum derivatives are produced in refineries, coal liquefaction or gasification processes demand the use of equipments and systems working in an environment involving high temperatures and pressures.

* Serious environmental issues are involved in going in for coal liquefaction and gasification facilities in a big way.

7

Pointers to Viability or Otherwise of Electricity Generating Power Plants

We had discussed on the various aspects of deriving energy from sources that are widely used for a long time, as well as the new ones we intend to bring into use in a big way in future. We have also found the importance of factors such as intensity of energy available from energy sources and the methods used by us to harness energy from various sources in the manner in which we are able to use them.

It is appropriate that we bring in more clarity on the influence of the various factors which determine the viability of energy conversion processes. The absence of this sort of evaluation could lead us to infructuous work in the energy conversion and utilisation area.

This seems to be evident from the manner in which we are pursuing with the capacity augmentation of on grid and off grid solar electric power plants in our country. The fact that we get some quantum of electricity out of these plants seem to make us believe that they are electricity generating power plants. By going in for objective evaluation of the related aspects, we will have clarity on the usability of these energy conversion processes in a viable manner.

In the following chapters, we are going in for an analysis of the factors related to the viability of the energy conversion systems used for producing electricity. These factors are:

(1) Cost per unit of electricity generated from different energy sources.

(2) Net energy available from each of the energy conversion systems used for generating electricity.

(3) The expenses incurred for producing 157.68 million electrical units during 30 year period from power plants working by drawing energy from different energy sources. This is the output per MW of electricity generating plants for a plant load factor of 60 per cent, and operation for 30 years

(4) Final conclusions on pointers to viability of elecricity generating power plants.

Cost Per Unit of Electricity Generated from Different Energy Sources

1. Sources of Energy for Electricity Generation

We generate electricity from coal, petroleum distillates, fission based nuclear energy, flowing water and wind. Until a few decades ago, the criteria for our choice of the type of power plant were largely a combination of local factors and the availability of appropriate electricity generating units. Concerns of large scale environmental pollution were a non-issue in those days.

Currently, generation of electricity from green energy sources is preferred. As a result of the same, harnessing wind energy has received a thrust during the past couple of decades or so. Governments are providing subsidies at various stages of establishment and operation of wind electric plants. This has motivated entrepreneurs to establish wind electricity generating farms at locations with favourable wind characteristics.

'On grid and off grid solar electric power plants' is the latest one in the green energy category we have decided to establish in a large scale. Incentives are provided for solar electricity generating units at all stages of establishment and utilisation, with the assumption that wide spread establishment of these could bring down the adverse effects of global warming. It has also been made mandatory that electricity distributors buy solar electricity from entrepreneurs at pre–decided rates (Ref: report titled, "Solar power tariffs being finalised" published in *The Hindu* of July 21, 2010).

According to this report the Kerala State Electricity Regulatory Commission (KSERC) defined solar power tariffs applicable to power distribution agencies buying electricity from roof top and small generating systems. The proposed tariffs are Rs. 16.31 a kWh for electricity generated by solar thermal electric units and Rs. 17.91 a kWh for solar photovoltaic units. This exercise is said to be in conformity with the Union Government's policy on development of solar electricity generation as a part of the Jawaharlal Nehru National Solar Mission.

Solar electricity generation is new for domestic and industrial consumers alike. Hitherto, these were used for remote areas wherein electricity supply grid was unavailable. In such cases the high cost of electricity generated was a matter of secondary concern. However, when we establish solar electric units as well as the conventional plants side by side, the cost per unit of electricity available need be compared.

Considering the above, the cost per unit of electricity generated from different energy sources is worked out.

2. Factors used for Working out the Cost per unit of Electricity Generated are given below:

(i) *Start up cost*

The establishment of an electricity generating power plant involves:

- Procurement of land and provision of infrastructure.
- Design, manufacture and installation of plant equipments.

These call for one time investment and is supported by financial institutions. Cost of capital consists of repayment of the principal as well as interest.

The start up cost of different types of power plants is tabulated below :

Table

Sl.No.	*Type of power plant*	*Start up cost/MW (Rs in millions)*
1.	Fission based nuclear electric	100.0
2.	Hydro electric	50.0
3.	Wind electric	70.0
4.	Thermo electric (coal based)	50.0
5.	Solar photo voltaic/solar thermal	200.0

(ii) *Life of power plant*

In general, life of power plants is assumed as 30 years. A 15 years life has been assumed for solar PV systems.

(iii) *Plant load factor*

Plant load factor indicates the average duration a power plant functions during a day. Theoretically, plant load factor of new plants belonging to thermoelectric, hydroelectric, nuclear electric categories could be close to 100 per cent. However, for various reasons, they work at much lower levels and therefore, their average plant load factor for life is assumed as 60 per cent. We are adopting 25 per cent for wind electric and 15 per cent for solar electric power plants.

(iv) *Expenses incurred for electricity generation*

Expenses incurred on a plant during its establishment and operation for life are:

(a) Capital is assumed to be returned over its lifetime.

(b) Interest on capital is assumed as 10.0 per cent.

(c) Yearly operational expenses are assumed as 3.33 per cent of the startup cost.

(d) Fuel cost.

(e) Distribution charges are assumed to be Rs. 2.0 million/MW for power plants with 60 per cent P.L.F. For lower plant load factors, this is proportionately brought down.

3. Calculation of Cost per unit of Electricity Generated Using Different Types of Power Plants

3.1. Hydro electric power plant

(Calculations based on 30 years life term and 60 per cent P.L.F. of the power plant/MW)

(a) Start up cost: Rs. 50.0 million.

(b) Life of power plant: 30 years.

(c) Plant load factor : 60 per cent.

(d) Cost of repayment on start up cost (principal + interest for 30 years period) = Rs. 125.0 million.

(e) Fuel cost : Nil.

(f) Operational cost for life (excluding fuel cost) : Rs. 50.0 million.

(g) Distribution charges for life: Rs. 60.0 million.

(h) Expenditure over 30 years period (d+ e + f + g): Rs. 235.0 million.

(i) Profit at 10 per cent of expenditure : Rs. 23.5 million.

(j) Expenditure for 30 years including profit: Rs. 258.5 million.

(k) Plant working hours per day (averaged over 30 years) : 14.4 hour.

(l) Output over 30 year period : 157.68 million unit.

(m) Cost of electricity generated : $\frac{\text{Rs } 258.5 \text{ million}}{157.68 \text{ million unit}}$

: Rs. 1.63 per unit.

3.2. Thermoelectric power plant
(Coal based)

(Calculations based on 30 years life term and 60 per cent P.L.F. of the power plant/MW)

(a) Start up cost : Rs. 50.0 million.

(b) Life of power plant: 30 years.

(c) Plant load factor: 60 per cent.

(d) Cost of repayment on start up cost (principal + interest for 30 years period) : Rs. 125.0 million.

(e) Fuel cost for life: Rs. 522.0 million.

(f) Operational cost (excluding fuel cost) for life : Rs. 50.0 million.

(g) Distribution charges for life : Rs. 60.0 million.

(h) Expenditure over 30 years period (d+e+f+g) : Rs. 757 million.

(i) Profit at 10 per cent of expenses: Rs. 75.7 million.

(j) Expenditure over 30 years including profit: Rs. 832.7 million.

(k) Plant working hours per day (average over 30 years): 14.4 hour.

(l) Output over 30 years period: 157.68 million units.

(m) Cost of electricity generated: $\frac{\text{Rs. 832.7 million.}}{\text{157.68 million units}}$

: Rs. 5.27 per unit.

Note: Annual requirement for coal is taken as 5.8 x 10^6 kg/MW. This figure is adopted from a media report published some time ago. Price of coal is taken as $60 per tonne; based on the then price of imported coal. This works out to Rs. 3.0 per kg.

3.3. Fission based nuclear electric power plant

(Calculations based on 30 years life term and 60 per cent P.L.F. of the power plant/MW)

(a) Start up cost : Rs. 100.0 million

(b) Life of power plant: 30 years

(c) Plant load factor: 60 per cent

(d) Cost of repayment on start up cost (principal + interest for 30 years period): Rs. 250.0 million.

(e) Fuel cost for life: Rs. 261.0 million

(f) Operational expenses (excluding fuel cost) for life: Rs. 100.0 million

(g) Distribution charges for life: Rs. 60.0 million.

(h) Expenses over 30 years period (d+e+f+g) : Rs. 671.0 million.

(i) Profit at 10 per cent : Rs. 67.1 million.

(j) Expenditure over 30 years including profit: Rs. 738.1 million.

(k) Plant working hours per day (averaged over 30 years) : 14.4 hour.

(l) Output over 30 year period : 157.68 million unit.

(m) Cost of electricity generated : $\frac{\text{Rs. 738.1 million}}{\text{157.68 million units}}$

: Rs. 4.67 per unit.

Note: Fuel cost is assumed as half the fuel cost incurred for coal based power plants.

3.4. *Wind electric power plant*

(Calculations based on 30 years life term and 25 per cent P.L.F. of the power plant/MW)

(a) Start up cost : Rs. 70.0 million.

(b) Life of power plant: 30 years

(c) Plant load factor: 25 per cent.

(d) Cost of repayment on start up cost (principal + interest for 30 years period): Rs. 175.0 million

(e) Fuel cost : Nil

(f) Operational expenses (excluding fuel cost) for life: Rs. 70.0 million.

(g) Distribution charges for life: Rs. 25.0 million.

(h) Expenditure over 30 years period (d+e+f+g): Rs. 270.0 million.

(i) Profit at 10 per cent : Rs. 27.0 million.

(j) Expenditure over 30 years including profit: Rs. 297.0 million.

(k) Plant working hours per day averaged over 30 years : 6.

(l) Output over 30 years period : 65.7 million units.

(m) Cost of electricity generated : $\frac{\text{Rs. 297.0 million}}{\text{65.7 million units}}$

: Rs. 4.52 per unit.

3.5. Solar electric power plant

(Calculations based on 15 years life term and 15 per cent P.L.F. for the power plant/MW)

(a) Start up cost : Rs. 200.0 million.

(b) Life of power plant : 15 years.

(c) Plant load factor : 15 per cent.

(d) Cost of repayment on start up cost (principal + interest for 15 years) : Rs. 350.0 million.

(e) Fuel cost: Nil.

(f) Operational expenses (excluding fuel cost) for life: Rs. 100.0 million.

(g) Distribution charges for life: Rs. 7.5 million.

(h) Expenditure over 15 years period (d+e+f+g): Rs. 457.5 million.

(i) Profit at 10 per cent : Rs. 45.75 million.

(j) Expenditure over 15 years including profit: Rs. 503.25 million.

(k) Plant working hours per day averaged over 15 years : 3.6.

(l) Output over 15 years period : 19.65 million units.

(m) Cost of electricity generated: Rs 25.5 per unit.

Note: Refer report titled "Solar power tariffs being finalised" quoted in this book. It could be seen that the price per unit of electricity is being finalised as Rs. 16.31 to 17.91. This is after giving concessions and subsidies at many stages of solar electricity generation. To arrive at the real cost, these motivators also have to be accounted and then the tariff could be in the range we have calculated here from first principles. In addition to providing subsidies at various stages of establishment and operation of solar electric plants, the government provides heavy subsidies on the cost of electricity generated also.

4. Conclusions

The cost per unit of electricity generated by the different types of power plants as worked out in this chapter is given in the following table.

Table

Type of power plant	*Cost per unit of electricity generated (Rs.)*
Thermoelectric	5.27
Fission based nuclear electric	4.67
Hydroelectric	1.63
Wind electric	4.52
Solar electric	25.5

Net Energy Analysis

Net energy analysis refers to the studies on the quantum of electricity generated by a power plant during its lifetime, in relation to the quantum of external energy required to construct, fuel and operate the same. This is important, since, a major portion of the energy input into the manufacture and life time operation of power plants is in the form of electricity.

Assessment of net energy available from electricity generating power plants is not usually dealt with in text books. In order that we proceed with our discussions, a study conducted on behalf of The Institute of Energy Analysis, USA on light water nuclear reactors of 1000 MW rating is referred here. The relevant portion of the report is given as Annexure 2.

It is essential that we are aware of the extent of net energy available from different types of power plants, since, this aspect clearly tells us about their viability for electricity generation. We shall therefore look at them one by one.

1. Method adopted for the calculation of net energy available from power plants functioning by drawing energy from different energy sources

The cost of electricity generated from various energy sources could guide us in our discussions on net energy analysis. We shall look at the net energy available from the different modes of electricity generation that are in use using relationships given below:

(1) The per cent of external energy input to a power plant of type 'X'

$$= \frac{\text{The per cent of external energy input to a nuclear electric plant} \times \text{Cost per unit of electricity generated from nuclear electric plant}}{\text{Cost per unit of electricity generated from nuclear electric plant}}$$

(2) Net energy available from plant 'X' = 100 per cent - per cent of external energy input to plant 'X'.

(3) Number of identical plants producible using the net energy available

$$= \frac{\text{Net energy available from plant 'X'}}{\text{External energy input for the manufacture and lifetime operation of one unit of X}}$$

2. Calculation of net energy available during the energy conversion processes from different types of electricity generating power plants

2.1. Fission based nuclear electric power plant

This study is on light water nuclear reactors. Major factors considered in the study included 30 years plant life and 61 per cent average plant load factor. The fuel was enriched by 0.205.

Calculation of net energy

The study concluded that for every 100 units of electricity output by the plant, the external energy input for its manufacture, fuelling and operation was in the range of 20 to 25 units. Averaging to 22.5 units, we get a net energy of 77.5 units.

If we utilise this net energy solely for the manufacture and lifetime operation of new plants of the same category, the number of similar plants we could make

$$= \frac{77.5}{22.5} = 3.4$$

Conclusion: The above result indicates that utilising the electricity generated by the plant, we could manufacture a replica of the plant. In addition, we get electricity required for manufacturing 3.4 similar plants. This is the net electricity from the plant which could cater to the electrical loads of domestic as well as industrial consumers.

This indicates that fission based nuclear electric plants currently in use are viable electricity production units.

2.2. Thermoelectric power plant

Thermoelectric plants produce the major portion of electricity in our country. The characteristics that could influence the net energy available from these include:

- Start up cost: Rs. 50.0 million/MW.
- 30 years plant life.
- Plant load factor averaged over life is 60 per cent.
- Input energy is from hydrocarbon based fuels. Fuel is a major component of the external energy input for the functioning of these plants.

A report which appeared in *The Hindu Business Line* of November 8, 2009 is referred here. The title of the report was, "Generation at key NTPC stations falters." The report highlighted about delays that occured in the import of coal. Kahalgaon and Farakka units were among the thermal projects which were forced to operate at 50 per cent plant load factor (PLF) due to delay in fuel imports. 1,357 million units was the loss of generation during September alone due to imported fuel failing to come through. The 2,340 MW Kahalgaon and 1600 MW Farakka units operated at a PLF of 45.4 per cent and 51.7 per cent respectively.

Further, the report said that during the year the coal requirement was around 150 million tonne. Since imported coal has high calorific value, 12.50 Mt of imported dry fuel is equivalent to around 19.0 Mt of domestic coal.

The report highlights the importance of maintaining high plant load factor for coal based power plants.

For working out the net energy available from thermoelectric plant, we have the following data:

- Cost per unit of electricity generated from thermoelectric plant is Rs. 5.27.

- Cost per unit of electricity generated from nuclear electric plant is Rs. 4.67.

- External energy input for the manufacture and lifetime operation of nuclear electric plant is 22.5 units out of 100 units produced (22.5 per cent).

Calculation of net energy

(1) The percentage of external energy input for the manufacture and lifetime operation of a thermoelectric plant

$$= \frac{\text{Per cent of external energy input into the manufacture and life time operation of a nuclear plant} \times \text{Cost per unit of electricity generated from thermo electric plant}}{\text{Cost per unit of electricity generated from nuclear electric plant}}$$

$$= \frac{22.5 \text{ per cent} \times \text{Rs } 5.27}{\text{Rs } 4.67} = 25.43 \text{ per cent.}$$

(2) Net energy available from thermoelectric plant for every 100 units of output = 100 – 25.43 = 74.57 per cent

(3) Number of identical plants producible using the net energy available

$$= \frac{74.57\ \%}{25.43\ \%} = 2.93$$

Conclusion: The above result indicates that by utilising the electricity available from the plant, we could manufacture a replica of the plant. In addition, we get electricity for manufacturing 3.4 more similar plants. This is the net electricity available from the plant which could be used for catering to the electric loads of domestic as well as industrial consumers.

This indicates that thermoelectric plants are viable electricity producing units.

2.3. Hydroelectric power plant

The following factors influence the availability of net energy from these plants:

- Start up cost is around Rs. 50.0 million/MW.
- Life of dams is over 100 years. However, plant life is taken as 30 years for the purpose of our calculation.
- No fuel is involved in its functioning.
- Plant load factor averaged over life is 60 per cent.

The cost per unit of electricity generated from a hydroelectric plant is worked out as Rs. 1.63. Utilising the same, we could calculate the net energy available from this type of plants as given below:

Calculation of net energy

(1) The per cent of external energy input for the manufacture and lifetime operation of a hydroelectric power plant

$$= \frac{\text{The per cent of external energy input to a nuclear electric plant} \times \text{Cost per unit of electricity generated from thermo electric plant}}{\text{Cost per unit of electricity generated from nuclear electric plant}}$$

$$= \frac{22.5 \text{ per cent} \times \text{Rs } 1.63}{\text{Rs } 4.67} = 7.86 \text{ per cent.}$$

(2) Net energy available from hydroelectric plant for every 100 units of output = 100 per cent − 7.86 per cent= 92.14 per cent

(3) Number of identical plants producible using the net energy available

$$= \frac{92.14 \text{ per cent}}{7.86 \text{ per cent}} = 11.7$$

Conclusion: The above result indicates that by utilising the electricity available from the plant we could manufacture a replica of the plant. In addition, we get electricity required to manufacture 11.7 more similar plants. This is the net electricity available from the plant which could be used to cater to the electric loads of domestic as well as industrial consumers.

This also indicates that hydroelectric plants are the most viable electricity production units.

2.4. Wind electric power plant

The following factors influence the net energy available from these plants:

- The start up cost is around Rs. 70.0 million/MW.
- Life of the plants is 30 years
- Plant load factor is generally low, and is in the range of 26 to 30 per cent. Average factor for life time of the units is taken as 25 per cent for our analysis.
- No fuel is involved in their operation.
- The cost of electricity generated is Rs. 4.64.

Calculation of net energy

(1) External energy input into the manufacture and lifetime operation of wind electric plants

$$= \frac{\text{Per cent of external energy input into the manufacture and life time operation of a nuclear plants} \times \text{Cost per unit of electricity generated from wind electric plants}}{\text{Cost per unit of electricity generated from nuclear electric plants}}$$

$$= \frac{22.5 \text{ per cent} \times \text{Rs } 4.64}{\text{Rs } 4.67} = 22.4 \text{ per cent.}$$

(2) The net energy that is available from wind electric power plants for every 100 units output = 100 per cent - 22.4 per cent = 77.6 per cent

(3) Number of identical plants producible using the net energy available

$$= \frac{77.6 \text{ per cent}}{22.4 \text{ per cent}} = 3.47 \text{ units.}$$

Conclusion: The above result indicates that by utilising the electricity generated by the plant we could manufacture a replica of the plant. In addition, we get electricity required for the manufacture of 3.47 more similar plants. This is the net electricity we receive from the plant, and could be used to cater to the electrical loads of domestic as well as industrial consumers.

This also indicates that wind electric plants built at carefully chosen sites are viable electricity production units.

2.5. Solar electric power plant

The following factors influence the net energy available from these.

- Start up cost of Rs. 200 million/MW.

- According to a recent report, permission is given by the Ministry of New and Renewable Energy sources to depreciate solar electric plants in seven years period. However, we assume a higher life of 15 years for our calculations.

- In addition, their plant load factor at best 15 per cent. This brings down the output energy in terms of electrical units to $1/4^{th}$ in comparison to the output from thermal/hydro/nuclear.

- The cost of electricity from solar electric power plant as Rs. 25.5.

Calculation of net energy

(1) The external energy that is input for the building and operation of solar electric plants for their lifetime :

$$\frac{\text{Rs. } 25.5 \times 22.5\%}{\text{Rs. } 4.66} = 117.5\ \%$$

(2) The net energy available from solar electric plants for every 100 units output

= 100 per cent - 117.5 per cent = –17.5 per cent

(3) Number of identical plants producible using the net energy available: Since the net energy available from solar electric plants is negative, number of plants producible using the net energy available is 'NIL'

Conclusion: Considering the above, we conclude that setting up grid connected solar electric plants utilising the current level of technology does not in any way contribute to the production of electricity, and hence, these do not function as electricity generating plants. Also, since these units are manufactured largely using electricity generated by thermal power plants and therefore, the establishment and utilisation of solar electric plants do not in any way contribute to mitigating global warming related dangers.

We could also conclude that the input energy for the manufacture of solar electric power plants is largely from sources other than solar electricity and hence we seem to be blind with respect to the information that no net energy is available from solar electric power plants.

2.6. Geothermal electric plant

There are many factors which are against this source of energy becoming viable for electricity generation.

Some of them are:

- Geothermal wells providing quality steam are rare.
- The efficiency of electricity generating units operated using steam from geothermal wells is in single digits only.

- Each well created for extracting geothermal energy could operate turbines of a fraction of MW capacity only. This small capacity of individual units adversely affects their viability.

- Creating steam production wells involve high capital cost. Wells are generally created by drilling bores in rock to a depth a few km, and then fracturing the rock at the bottom using explosives. It is this fractured portion which is used for geothermal heat extraction from hot rocks. For the successful fracturing of the rocks, use of atom bombs also has been suggested. We know that atom bombs are neither cheap nor easily available. This aspect indicates the difficulties encountered in fracturing the rocks at the bottom of a bore which is at a depth of a few kilometers from the surface of earth.

Therefore, the possibility of geothermal energy source becoming viable for electricity generation is remote.

2.7. Ocean wave/tide – electric plant

Ocean wave/tide is considered as a green energy source for electricity generation. Studies have been conducted by institutions by setting up experimental units. However, as we already know, waves/tides have characteristics that could work against the viability of electricity generation units.

They are:

- Waves are of intermittent nature.

- They contain low level of energy per unit volume of sea water.

Some details of a study conducted by IIT, Chennai on a plant established at Vizhinjam, near Thiruvananthapuram, as well as a plant at Portugal have been quoted in this book. Limited information available through media reports on these suggest that this source might not be viable for electricity generation.

3. What if the threat to global warming turns out to be a farce?

Nowadays, serious debates are going on about the disaster that is feared to come on us due to the man made global warming. However, there is a section among us who believe that the issue of global warming is farce.

The seemingly threatening situation of global warming is making us to pursue the following:

- Work on energy saving methods and the create awareness on the need to conserve energy.
- Implement design improvements with added stress on energy efficiency.
- Research on deriving energy from alternate sources belonging to green energy category.

We could see the above as long term benefits of the ongoing debate on global warming. We have a win-win situation here.

Actions taken by us in this regard could also be beneficial on the following counts:

- Reductions in the consumption of fuels will extent the life of their reserves. A few more generations coming after us could enjoy benefits of using the non-renewable as well as high intensity energy sources, namely coal and petroleum.
- In case we are able to produce electricity from new green energy sources in a viable manner, we will pollute our environment to a lesser extent. Generations after us might have a cleaner environment to live.

4. Conclusions

(1) The net energy available from different types of power plants and the number of similar plants producible utilising solely the net energy output are given in the following table.

Table

Type of power plant	*Net energy produced (out of each 100 unit)*	*Number of similar plants producible by utilising output energy*
Nuclear electric	77.5	3.4
Hydroelectric	92.14	11.7
Thermoelectric	74.57	2.93
Wind electric	77.6	3.47
Solar electric	Nil	Nil

(2) A rough estimate on the net electricity available from the various types of power plants has been carried out.

The study indicates the following:

* Well designed hydroelectric power plants stand at the top of the table with respect to the net electricity available.

* Next on the table we have thermal, fission based nuclear and wind electric power plants. Significant quantum of net energy available from them indicates their viability for electricity generation.

* Solar electric power plants belong to two distinctly different categories, namely, solar photo voltaic and solar thermal. Studies indicate that no net energy is available from solar electric plants built utilising the present level of technology and hence, they can not be considered as belonging to the category of electricity generating power plants.

* The case of plants powered by ocean waves/tides and geothermal seems to be no different either. These energy

sources are of very low intensity in comparison to the energy intensity available from fuels of biological origin. Therefore viabilities of electricity generating power plants seem to be doubtful.

Kind attention please....

More than the accuracy.of the data used for the calculations and that of the result obtained, this exercise is carried out to bring in clarity on the various aspects related to the functioning of electricity generating power plants, particularly addressing to their viability aspects for electricity generation. The manner in which today's energy managers and administrators deal with energy supply related issues seems to indicate that they do not go in for in depth study on such aspects. This is clearly visible from the manner in which capacity additions of solar electricity generation units are being promoted in our country. It is generally pointed out by them that the deficiency in solar electricity generating systems lies only in their high cost; and for making them affordable governments need to provide subsidies. However, no one seems to have bothered to look at the disease beneath the symptom of high cost as well as other negative aspects of solar electricity generation. The establishment solar electricity generation systems involves huge expenditure, and we have already seen that the benefits are not commensurate with the expenses incurred.

Expenses for Producing 157.68 million Units in 30 years from Power Plants

1. Reason behind discussing the production of 157.68 million units in 30 years.

We have seen in earlier chapters the following regarding conventional electricity generating plants:

- Conventional electricity generating plants we have been using for a long time consist of hydroelectric, thermoelectric and nuclear electric ones. All these units could be operated at high plant load factor of even 100 per cent on emergent situations. However, in practice, its average value is much lower. For our analysis, we have assumed the plant load factor of these plants as 60 per cent.

- Life of the above power plants is assumed as 30 years.
- Electricity generating power plants of 1 MW installed capacity belonging to the above categories generate 157.68 million units of electricity over their life span of 30 years at a plant load factor of 60 per cent.
- The departure from this situation could be seen in the cases of wind electric as well as solar electric power plants. The plant load factors of wind electric plants are assumed as 25 per cent, and solar electric plants as 15 per cent.

2. Data worked out earlier

In the previous chapters we have discussed about hydroelectric, thermoelectric, nuclear electric, wind electric, solar electric power plants.

Data worked out on expenses incurred for 1 MW power plants on during their life time are given below:

Serial No.	*Type of power plant*	*Life (year)*	*Plant load factor (%)*	*Expenses for 30 year/MW (million rupee)*
1.	Hydroelectric	30	60	258.5
2.	Thermoelectric	30	60	832.7
3.	Nuclear electric	30	60	738.1
4.	Wind electric	30	25	297.0
5.	Solar electric	15	15	503.25

3. Working out expenses for producing 157.68 million units in 30 years from power plants.

3.1. Expenses for producing 157.68 million units during 30 year period from hydro/thermo/nuclear power plants

These plants have 30 years life, and work at an average PLF of 60 per cent. In addition, they produce around 157.68 million units of electricity annually per MW of installed capacity, and therefore, the expenses are (1) Rs. 258.5 million for hydroelectric plant, (2) 832.7 for thermoelectric, and (3) Rs. 738.1 million for nuclear electricplant. These figures are already available in the table.

3.2. Expenses for producing 157.68 million units during 30 year period from wind electric and solar electric power plants

3.2.1. Wind electric

Since these plants are considered to work at a plant load factor of 25 per cent over their life, in order that we get an output of 157.68 million units over 30 years period, their installed capacities have to be augmented as

$$\frac{1 \text{ MW x 60 per cent}}{25 \text{ per cent}} = 2.4 \text{ MW.}$$

Correspondingly the expenses for producing 157.68 million units out of wind electric plants goes up to

$$\frac{2.4 \text{ MW x Rs. 297 million}}{\text{MW}} = \text{Rs. 712.8 million.}$$

3.2.2. Solar electric

These plants need upgradation with respect to two different aspects.

(a) The first one is with respect to their low plant load factor of 15 per cent. This demands that the installed capacity has to be upgraded to

$$\frac{\text{MW x 60 \%}}{15\%} = 4 \text{ MW.}$$

Therefore, expenses incurred on installing and running solar electric plant of 4 MW installed capacity during their life time of 15 years is

= 4.0 MW x Rs. 503.25 million

= Rs. 2013.0 million.

(b) In addition, since the life of solar electric plants is considered as 15 years; they need to be established twice during the 30 years period.

Therefore, the expenditure incurred in establishing and running solar electric plants twice during 30 years period = Rs. 2013 million x 2 = Rs. 4026 million.

4. The expenses for producing 157.68 million units from the power plants over 30 years period are tabulated below.

Serial No.	*Type of Power plant*	*Expenses/MW of installed capacity for life. (Rs. in millions)*	*Expenses for generating 157.68 million units over 30 year period. (Rs. in millions)*
1.	Hydroelectric	258.5	258.5
2.	Thermoelectric	832.7	832.7
3.	Nuclear electric	738.1	738.1
4.	Wind elecric	297.0	712.8
5.	Solar electric	503.25	4026.0

Gist of Pointers to Viability of Electricity Generating Plants

1. **Table below gives brief on pointers to the viability/non viability of electricity generating plants.**

Type of power plant	*Cost per unit of electricity generated (Rs.)*	*Number of similar plants producible from net energy*	*Expenses for generating 157.68 million units over 30 years (Rs. in millions)*	*Remarks*
Hydro-electric	1.63	11.7	258.5	Expenses are the lowest, hence most viable
Thermo-electric	5.27	2.93	832.7	Viable
Nuclear electric	4.67	3.4	738.1	Viable
Wind electric	4.52	3.47	712.8	Viable
Solar	25.5	nil	4026.0	(1) Expenses are 15.4 times the lowest (2) No net energy Produced; hence points to non viability

2. The differences in energy intensities:

(a) The energy intensity available from the fire created by burning fuels is the highest. There is no energy source available with energy intensity which is obtained by the burning of fuels.

(b) For meeting our energy needs of the past in the areas heat/ electricity generation, we used energy forms with high intensities. They consist of fuels, hydro power and wind power of high intensities and energy from nuclear fusion.

(c) The new energy sources we seem to believe as our saviours in the coming times, namely solar energy, geothermal energy, energy from waves/tides are of very low intensity compared to the energy intensity available from fire.

Note: The higher the intensity of source we use for electricity generation, chances are that the plants are viable, and lower the intensity of the source they move to non-viable category.

3. Studies on net energy available are a must before adopting an energy conversion process for use in a large scale.

The study conducted by using the net energy available from different types of power plants and the conclusions on their viability aspects are given below:

3.1. Viable electricity generating plants

3.1.1. Hydroelectric power plant: Hydroelectric power plants provide 157.68 million units on an average over 30 years period at a lifetime expense of Rs. 258.5 million per MW of installed capacity and the price per electrical unit produced is worked out as Rs. 1.63. Our caculations indicate that 11.7 similar plants could be produced from the net energy available from a plant. This energy in the form of electricity is available for our varied applications.

3.1.2. Wind electric power plant: Next in line comes wind electric plants producing 157.68 million units at a lifetime expense of Rs. 712.8 million over 30 years period and the cost per unit of electricity produced is worked out as Rs. 4.52. The net energy could produce 3.47 similar plants and this could be used for varied applications.

3.1.3. Fission based nuclear electric power plant: Then we have the nuclear electric plants producing 157.68 million units of electricity over 30 years period at a lifetime expense of Rs. 738.1 million, and the cost per unit of electricity produced is worked out as Rs. 4.66. The net energy available from the plant could be utilised to produce 3.4 similar plants, and this could be used for varied applications.

3.1.4. Thermoelectric power plants: Thermoelectric plants produce 157.68 million units over a 30 years period at a lifetime expense of Rs. 832.7 million, and at a cost per unit of electricity produced is worked out as Rs. 5.27. The net energy available from the plants could produce 3.4 similar plants, and this could be used for varied applications.

3.2. Non viable electricity generating plant

3.2.1. Solar electric power plant : Solar electric plants produce 157.68 million units of electricity over 30 years period at a life time expense of Rs. 4026 million, and at a cost per unit of electricity produced is worked out as Rs. 25.5. According to this study, since more energy has to be input for the manufacture and lifetime operation of solar electric plants compared to the energy available from the same during their life time, no net energy is available from these. Due to the same reason, our studies indicate that they do not function as electricity generating plants.

4. Factors contributing to viability/non-viability of energy conversion systems such as electricity generating plants, fuel cells, bio fuel production.

- High intensity of input energy (e.g., fuels) contributes to viable energy conversion/electricity generation. Low intensity (e.g., solar energy, wave energy, geothermal energy) could contribute to non-viability.
- High plant load factor could contribute to viable energy conversion, and evidently, substantially low plant load factor (e.g., solar electricity generation, wave energy harnessing) could contribute to non-viability.
- Single units with high capacity contribute to viability; the integration of tiny units to make large units (solar electricity generating units in which bigger units have to be made from basic cells of 0.49 W) contribute to nonviability.

- Low plant life (e.g., solar electricity generating units, fuel cells) contribute to non-viability.

- Extensiveness of power plants contributes to non-viability (e.g., solar electricity generating units).

The need to upgrade low voltage (0.45 V) electricity generated by solar cells into grid compatible high voltages contribute to their non-viability.

8

Alternatives to Athirapally Hydroelectric Project

1. Sanction for Athirapally hydroelectric project is reconsidered

Kerala State intends to setup a hydoelectric power station at Athirapally, situated in the high ranges of Kerala. The proposal is to have four units of around 40 MW each; adding to a total of 163 MW. However, some agencies have differences of opinion on setting up this station. Their concerns are largely based on the following:

(1) Bio diversity of the area could be affected and hence, a detailed study regarding the same need be carried out.

(2) Adivasi families need be rehabilitated.

It is possible that by the time this book is published, the project gets green signal for implementation. Even under such circumstances, the analysis presented here would help us in understanding the significant differences among electricity generating power plants functioning by drawing energy from different sources.

2. Analysis

Deepa: I am hearing a lot about the hitches in the implementation of a hydroelectric project at Athirapally, located in the State of Kerala. Recently, it was reported that its sanction for implementation is being withdrawn by the Government of India (report January 2010).

Devarajan: Why this reconsideration? As I understand, these are times when we are putting all out efforts to increase the installed capacity of power plants belonging to green energy category. Hydroelectricity ranks no. 1 among all the non-conventional energy sources for the production of electricity.

Lekshmi: That is true. If we look from another angle, we are putting serious efforts for generation of electricity from one of the weakest energy source, namely, the solar rays. We are pushing forward a seemingly ambitious plan of augmenting the installed capacity of solar electric plants. Our target is to reach 20000 MW, by the year 2022.

2.1. History of hydroelectricity generation in Kerala

Devarajan: At this juncture it would be appropriate to go through the history of development of hydroelectric plants in our state. It will help us to develop a better understanding on this topic.

Muthu: As we are aware, a significant portion of Kerala is covered with mountains. When the question of electricity generation was considered over half a century ago, hydroelectricity was the natural choice.

Devarajan: Why did you mention it as a natural choice? I understand that most of India's power stations belong to thermoelectric category, and are fuelled by coal.

Muthu: Our neighbouring states are blessed with coal deposits, we are not. Hence the natural choice was hydroelectricity generation.

Lekshmi: A major capacity addition was that of Idukki hydroelectric project (780 MW) which was commissioned over three decades ago. Presently, this power station is the backbone of electricity generation in our state.

Nagaraj: Many wondered about our wisdom in going for such a large power station then.

Lekshmi: Our demand for electricity has been continuously rising and therefore, around two decades ago, NTPC established a 230 MW thermoelectricity generating unit at Kayamkulam. This was initially a coal fired unit which was subsequently converted into a combined cycle one. Barring this, and the diesel electricity generation facility of 120 MW at Brahmapuram, all the other units of Kerala are hydroelectric ones.

Nagaraj: It is therefore evident that Kerala is tuned well for hydoelectricity generation. In such a situation, why did our proposal for setting up Athirapally power project meet with stiff resistance from the concerned?

Lekshmi: We own a generation capacity of 80 per cent in this category. Therefore, I also feel that our proposal for setting up new projects such as the one at Athirapally must be technically sound. Let us identify the reasons for this hitch through a detailed discussion.

Deepa: I do not know as to whether we have the data to carry out such a detailed analysis. However, I have a question in my mind. Assume that the hydroelectric projects that are doing yeoman service to our state are not already put up; and we are seeking permission for setting them up now. How many of them would have got clearances is anybody's guess.

Devarajan: Earlier we had referred to our National Solar Mission. When we are seriously pursuing with harnessing electricity from very weak energy sources such as solar rays, why is it that we are not getting clearances for the setting up of a hydroelectric project such as Athirapally?

Muthu: I do not know. Perhaps it might be outside our capability to perceive the reasons. However, approach from common sense point of view might provide the probable answer.

Lekshmi: Let us try to come out with reasons. About half a century ago we had higher quantum of forest coverage. When we proposed to establish such power projects in regions with thick forest coverage, we did not feel the pinch then.

Deepa: It could be a valid point. Another could be that generation of electricity itself was at its infant stage many decades ago and therefore, whatever project was perceived as feasible was implemented. There were no national grids at that time, no national authorities to oversee.

Devarajan: Could be true. Land acquisition and resettlement of those affected by the implementation of hydroelectric projects were not serious issues. The perceived hurdles were only technical ones.

Nagaraj: We tend to find solutions to easy problems first. Athirapally might have been one of the last in the queue.

Deepa: At this point of discussions what is our conclusion on this issue?

Muthu: If Athirapally was taken up half a century ago, it would have been a cake walk. The project goes for reconsideration when more experts debate, and the negatives get more prominence than the positives. In addition, lack of single agency to look at the energy related issues in a comprehensive manner might be the cause for the current dilemma.

Lekshmi: Yes, at this stage the conclusion seems to be in order.

Devarajan: I do not think that we should discuss about withdrawal of sanction for Athirapally in isolation, especially since we claim to be putting serious efforts to increase the contribution of renewable energy sources for electricity generation.

2.2. Looking for alternatives to Athirapally

Lekshmi: Hydroelectricity should qualify for 100 per cent marks, there is no doubt. When we are withholding the implementation of such a platinum class renewable energy project, we should be clear about the alternatives.

2.2.1. Thermoelectric plant

Deepa: One alternative is that we install thermoelectric power plants. This has short gestation period, it could be installed at chosen locations.

Nagaraj: I suggest that we install a plant belonging to renewable and green energy category as alternative. This is especially so in our present situation, wherein, we are very serious about reducing the perceived adverse effects of global warming.

2.2.2. Wind electric plant

Lekshmi: Let us look at green energy sources. We could setup a wind electric plant having capability to generate the same quantum of electrical units as that of Athirapally. Since plant load factors to which a wind electric plant could function is substantially low compared to that of hydroelectric plants, the installed capacity of wind electric plant with equivalent output to is

$$= \frac{160 \text{ MW} \times 60\%}{384 \text{ MW}} = 25\%$$

However, we have to locate an appropriate site for their installation. A counter argument could be that in case of our locating an appropriate site for tapping wind electric potential, we will any way be doing it. Therefore, this cannot be considered as an alternative to a most viable hydroelectric project as Athirapally.

The above argument is also valid.

2.2.3. Fission based nuclear electric plant

Deepa: Another green alternative is to go in for fission based nuclear electric plants of equivalent capacity. Setting up of these involve long gestation period, higher start up capital, issues of radioactivity related safety etc. In spite of all these disadvantages, this could be a long term alternative.

2.2.4. Solar electric plant

Muthu: Since we are going ahead with the setting up of solar electric plants in a big way as a part of National Solar Mission – 2020, let us understand as to what sort of price we have to pay for this green alternative. Again, the capacity of solar units should be such that they produce the same quantum of electrical units as that of 160 MW installed capacity of Athirapally.

Lekshmi: Let us work it out. We will have to bring out all the related issues so that we make a genuine comparison. These aspects have been discussed elsewhere in this book. We are bringing out three factors for this comparison.

2.2.4.1. Low plant load factor for solar electric plant

First of all, let us look at the difference in plant load factors related to these. They are 60 per cent for hydroelectric plants and 15 per cent for solar electric ones.

The direct effect of the vast difference in plant load factors is evident in the installed capacity we will have to provide for solar electric plants for producing the same quantum of electrical units as a 160 MW hydroelectric plant. Since the plant load factor of solar electric plant is only one fourth of hydroelectric plant, for deriving the same electrical output, we will have to provide an installed capacity of $= \dfrac{160 \text{ MW} \times 60\%}{15\ \%} = 640 \text{ MW}$ of solar electric plants.

2.2.4.2. Low plant life for solar electric plant

Devarajan: I shall highlight the second point. This is with respect to the difference in the operational life of the two types of power plants. We know that hydroelectric power stations have long life, especially with respect to dam and other civil constructions. However, for our discussions, we shall assume the life of a hydroelectric plant as 30 years. In contrast, many sub systems of solar electric plants are accepted to have a short life span. According to a media report published some time ago, National Solar Mission has accepted depreciating the capital invested on solar electric plants in a short time span of seven years. This gives an indication on the life of these. However, we shall assume a higher depreciation period of 15 years for our discussions.

Deepa: Are you aware of the implications of short life of solar electric plants?

Devarajan: Sure, during the 30 years life of hydroelectric plants, solar electric plants have to be established afresh twice. In other words, they have to be totally replaced every 15 years.

Deepa: This will have a direct bearing on the start up capital required for these.

2.2.4.3. High startup cost for solar electric plant

Nagaraj: I shall present the third point. This brings out the difference in startup costs of electricity generating plants.

Muthu: Please go ahead.

Nagaraj: For large hydroelectric power plants, the current rate is around Rs. 50 million per MW of installed capacity. Since the units at Athirapally are of only 40 MW installed capacity, we shall use a startup cost of Rs. 70 million/MW.

Lekshmi: Then the startup cost for the 160 MW Athirapally project is Rs. 11,200 million.

Nagaraj: We assume a startup cost of Rs. 200 million/MW for solar electric plants.

Lekshmi: It means that the startup cost/MW of solar electric plants is around 3.5 times that for hydroelectric plants.

Deepa: I shall represent this data in the form of a table.

Table

Type of power plant	*Hydroelectric*	*Solar electric*
Plant load factor	60 per cent	15 per cent
Plant life	30 years	15 years
Start up cost	70 million/MW	200 million/MW

3. Implications in brief

Devarajan: In the chapter titled, 'Pointers to the viability of electricity generating plants', we have worked out the life time cost of different types of power plants. The figures are as given below:

Table

Serial No.	*Type of power plant*	*Cost for the production of 157.68 million units in 30 years (Rs. in million)*
1.	Hydroelectric plant	258.5
2.	Thermoelectric plant	832.7
3.	Nuclear electric plant	738.1
4.	Wind electric plant	712.8
5.	Solar electric plant	4026.0

Deepa: I can't believe this. How come we have landed into such an astronomical figure with respect to the life time cost for solar electricity generation?

Muthu: This is nothing new. We have discussed these aspects in the related chapters. Many data have been taken from reports published in the recent past.

Lekshmi: That is true. I have also seen many reports in our National Dailies while the National Solar Mission was in its formative stage a couple of years ago.

Muthu: The lifetime cost of a power plant has a direct bearing on its viability. On refering to the above table, it could be seen that the lifetime cost for a hydroelectric plant which could produce 157.68 million units of electricity in 30 years is Rs. 258.5 million, and is the lowest among all the electricity generating plants. It could also be seen that lifetime cost for solar electric plants which could produce the same quantum of electrical units in 30 years is worked out as Rs. 4026.0 million. This is over 15 times the lifetime cost of hydroelectric plant giving the equivalent output.

Devarajan: Then what is our conclusion on our pushing through the establishment of grid connected as well as non-grid connected solar electricity generating systems in a big way?

Deepa: I believe that our discussions have brought out the related aspects clearly. We have already seen that the cost involved in setting up solar electric plants using the present level of technology and operating them for life is astronomical. Naturally, this makes them non-viable. More on the viability aspects of electricity generating plants could be read from the relevant chapters of this book.

Lekshmi: Sometime ago I have seen reports on the establishment of hydroelectric projects in China. The report mentioned that more than a million people were affected by the same.

Devarajan: Implementation of hydroelectric projects involve loss of forest area, relocation of inhabitants and the like. Whether we call it environmental damage or inevitable environmental modification depends on one's outlook. One thing is for sure; electricity is produced from these in the most viable manner.

Nagaraj: I shall conclude now:

(1) A well designed hydroelectric power plant is the best eternal power source for electricity generation our coming generations can enjoy.

(2) Such precious energy sources are location specific and limited. Other types of power plants are bound to create higher level of environmental damage in some way or the other, than the hydroelectric ones.

(3) It might not be wise to shelve projects such as the one we intend to establish at Athirapally, especially in the present times, when we are believing that global warming related issues are at our door step. Moreover, presently, we are looking for generation of electricity even from weak energy sources such as solar electricity.

(This chapter and the one that follows, namely, "Energy conservation" are presented in the form of a discussion among a few individuals, since, these are extracts from my first book on this subject titled, "Energy scenario beyond 2100", published during November 2010. However, responsibility for the contents of these chapters lie solely with this author.)

4. Tail piece

We shall look at a familiar story to bring out a clarity on the dilemma we face on the present energy scenario.

X : What are you searching for? Did you drop down something?

Y : I dropped down my ear rings, and I am searching for the same.

X : Where did you drop? Could I help in the search?

Y : It was dropped at the far end corner.

X : Then why are you searching for here?

Y : Because light is only here.

The story ends.

Lekshmi: Why did you quote this story here?

Muthu: It is feared that we will damage bio diversity due to the setting up of 163 MW hydroelectric power project at Athirapally, in the Western Ghats of Kerala. What this author wants to point out is that Kerala has Western Ghats stretching all the way from south to north along its eastern boundary for a distance of around 600 km. These are regions of above average rain fall, and are covered with forests having different species of plants and animals. However, we have only few regions in Western Ghats, suitable for setting up viable hydroelectric stations.

This is the same case with hydroelectric power stations set up in our country. Therefore, I believe that we should look at this issue in totality.

This story is quoted here to indicate that we are searching in the dark with respect to harnessing energy from different energy sources, when evident and viable solutions are before us and are not adopted.

9

Energy Conservation

We are consuming energy in the forms of electricity and fuels. We shall debate on each of these in this chapter.

1. Electricity

Deepa: Nowadays, the price of electricity that is made available to us is on debate frequently, especially in the State of Kerala where the widely accepted principle is that the prices of essential commodities and services should be contained to affordable levels. Hence, whenever an upward revision in electricity tariff is proposed, concerns are raised by the general public.

Some of the reasons mentioned for the hike are as given below.

- Availability of water from monsoon has been deficient. As a result of the same, storage reserviors of hydroelectric power stations were only filled partially. Our electricity generating plants are largely hydroelectric.

- The demand for electricity is spiralling up resulting in a wide gap between generation and consumption. This necessitates that in order to meet the demand, our Electricity Board is compelled to procure electricity from Central

Electricity Supply Authority and also from private suppliers. The price charged by them is significantly higher, compared to the expenditure we incur for the generation of electricity from our own hydroelectric stations. We have already seen earlier that well designed hydroelectric plants produce electricity at the cheapest rates. It is also true that most of the plants supplying electricity from locations that are outside our state are coal based. Hence, their cost of production will be on the higher side.

Muthu: You have projected a genuine problem. This issue is not pertaining to the State of Kerala alone. There are very few states in our country who are generating electricity in quantities matching to their internal demand.

The reasons for the wide gap existing between supply and demand are many. Some of these are given below.

- Cities, towns and villages are growing at high rates creating high growth in demand for electricity too.
- A large variety of home appliances have become part of almost every home.
- The demand for electricity by manufacturing industries is continuously rising.
- The gestation period for new power plants could be in the range of 5 - 10 years, whereas manufacturing industries could be setup in shorter time periods.
- Often, many of the electricity generating power plants are operated at low plant load factors. This is so, since, many a times we face difficulties in providing coal in quantities required for fuelling thermoelectric power stations and uranium for fuelling nuclear electric power reactors. Moreover, we also face shortages in the availability of water stored in the

reservoirs of hydroelectric power stations due to the under performance of monsoon seasons. This has serious impact on the operating efficiency of these, and in turn, on the price consumers have to pay for electricity.

- Significant quantum of electricity produced is wasted on various counts. The transmission losses are reported to be on the higher side.
- According to reports appearing in the media often, thefts do take place in a large scale resulting in loss of revenue. This burden comes largely on consumers who promptly pay electricity charges.
- Many large industrial establishments default payments on the electricity they use, and in due course they become red.

Lekshmi: I believe that you have brought out only one side of the story. The other side is that eventhough people express concerns on the price increase and desire that the expenditure incured on this account is brought down, awareness in them regarding the related issues is generally low. The Electricity Regulatory Authority, Kerala wanted the consumers to understand the seriousness of the situation and therefore, they went in for extensive campaigns highlighting the various means to bring down the consumption. However, they succeeded only by a limited extent.

Muthu: A major section of the public lack information on the technicalities of generation of electricity, reasons for its shortage, and the dire need to reduce consumption.

Lekshmi: That is very true. In addition, some of the large institutions, upper class and upper middle class citizens rarely set an example by adopting economy measures. In short, role models are few and far in between. Role models could serve as examples in the society, public at large follow them.

Devarajan: Many belonging to upper middle class and upper class think that they can afford to pay for electricity at high rates and therefore, there is no need for them to cut down the consumption of electricity.

Muthu: That is correct. There is another side to this issue. Many do not pay for the electricity used by them, or rather wasted by them. This is the case with those working in public and private institutions. There is no ownership for the problems arising out of wasteful consumption by the employees of these. Very often, managements do not put in efforts to educate their employees on the related aspects, and thereby avoid the avoidable wastage of electricity. The slogan, 'energy saved is energy produced' must percolate down to all sections of our society, encompassing administrators and the administered in an equal measure.

Nagaraj: There seems to be conformity of opinion among energy administrators on this issue. When opinions differ, the general public is confused.

Devarajan: Considering the need for effecting energy conversion at all costs, there could be a need for downward revision of lighting requirements for various categories of utilities. In addition, methods could be devised for lighting only in need. In this regard, it is worthwhile to recall a procedure Indian railways has adopted towards saving electricity. They have implemented a novel method of restricting the lighting on railway platforms, only for a predetermined duration, in relation to the presence of train at the platform.

Deepa: I think that the electricity suppliers and administrators need to pay more attention to this area of energy savings, since, they are the agency responsible for managing the whole issue. It is also their responsibility to educate the public on such matters. I am sure that the general public at large will cooperate on such important issues. What the general public lack is probably information on the need for reducing consumption, as well as the consequences of wasting the precious energy source.

Lekshmi: Perhaps it is appropriate to mention here about the campaign against the use of tobacco. Cigarettes are sold at shops, at the same time, consumers are told about the harmful effects of smoking through media in many ways.

In a similar manner, along with each bill for electricity consumed, consumers could be reminded about the problems we face in this area, and the advantage that could be available to us on consuming electricity for needs only.

Muthu: I agree. Economy in the consumption of electricity has to be practiced by all sections of society, irrespective of one's income level. This is essential, considering that the reserves of hydrocarbon based fuels are depleting at higher and higher rates, and also the perceived threat of global warming.

Coal, Petroleum etc.

Devarajan: By debating on the need for conservation of electricity, we have addressed only a part of energy saving requirements. However, the need for conservation extends to all forms of non-conventional energy forms. Today, we are largely using these for generation of electricity; renewable portion constitutes a minor part only.

Muthu: O.K. By saying that the consumption of electricity must be need based, we have indirectly effected savings in the utilisation of hydrocarbon based fuel forms, namely, coal and petroleum to some extent. In order that further energy savings are brought in, we will have to address the scope for savings in the input to these plants.

Nagaraj: True. Energy intensity improvements we pursue have a direct bearing on the energy savings. By this, we carryout performance improvements of power plants, so that we need to supply only less and less quantum of fuel for every unit of electricity output. This aspect is addressed by those working in the area of power plant performance improvements.

Lekshmi: Another area requiring our attention is the consumption of petrol, diesel, aviation kerosene and the like. Their consumption could be cut down substantially by more and more of us tuning our minds to using public transportation systems.

Deepa: That is highly desirable. Any item we buy from market invariably has energy input into it. Being prudent will take us a long way in effecting savings in coal and petroleum and to the extent, the coming generations shall enjoy the benefits.

10

Green Energy Sources and Future Energy Scenario

1. Population

India's population as per census carried out at 1991, 2001 and 2011 are given in the table A. In addition, extrapolated figures up to 2101 at 10 years intervals are provided in table B.

Table A
(Data from Census India)

Year	*Population (in Crore)*	*Rate of growth at 10 year interval (%)*
1991	84.0	-
2001	102.0	21.4
2011	121.0	18.6

Table B
(Data Extrapolated by the Author)

Year	*Population (in Crore)*	*Rate of growth at 10 year interval (%)*
2021	139.6	15.0
2031	157.3	13.0
2041	174.6	11.0
2051	190.6	9.0

(Contd.)

(Contd. Table B)

Year	*Population (in Crore)*	*Rate of growth at 10 year interval (%)*
2061	203.6	7.0
2071	213.8	5.0
2081	221.6	3.0
2091	228.2	1.0
2101	228.2	0.0

The world population which touched 700 Crore in October 2011, is expected to be around 1010 Crore at around 2050, and 1170 Crore at 2100. This data is from the book titled "Global Energy Perspectives," Edited by Nebojsa Nakisenovic, Arnulf Grubler and Alan McDonald and published by Cambridge University Press in the year 1998.

It is predicted that the increase in world population in the coming decades will largely be in the less developed countries of the present, and India is one among them.

2. Population and energy utilisation are closely related

Human population that existed at any period of time and the energy that was available then for use are related. This is clear from the fact that our population touched one billion (100 crore) only around 1804 A.D. The population also used to dip in situations of war and natural calamities.

The mining, processing and utilisation of hydrocarbons in the form of coal and petroleum in a large scale, has made the world population to increase to very large numbers during the twentieth century and later, in comparison to that existed at any point of time in the history of mankind. In addition, the availability of this additional quantum of energy as mentioned above, has enabled the maintenance of good portion of human population at a higher level of comfort compared to the past.

3. Present Scenario of Energy Utilisation

* India's consumption of energy during the year 2010 - 11 was 477 million tonnes oil equivalent, according to a report in *The Hindu* of January 19, 2012. This quantity is inclusive of all the energy forms put together. Out of this, 45 per cent consisted of oil and gas put together, and 10 per cent was natural gas alone.

* The annual target for production of coal in our country for the financial year 2011 - 12 was 447 million tonnes, according to data given by the Chief of Coal India and reported in *The Hindu* of January 23, 2012.

* In the area of electricity generation, our country's installed capacity is reported to be 173,600 MW at March 2011. This consists of a green energy component of around 55,000 MW (mostly hydro and wind), and the thermoelectricity part of 118,600 MW.

* *The Hindu* of October 16, 2012 published a report titled, "Bangladesh keen to join TAPI project". The report which was based on a statement by the Petroleum and Natural Gas Minister, Government of India, Shri Jaipal Reddy contained information on the import of crude by India during 2011 - 12 as given below:

- In 2011–12, Saudi Arabia supplied 32.3 million tonnes, Iraq supplied 24.51 million tonnes, Kuwait supplied 17.67 million tonnes, and Iran supplied 17.10 million tonnes. For 2012–13, imports from Iran is expected to be lower.

(The report is in the background of mandates to India to reduce crude oil imports from Iran as a part of sanctions. Sum total of the above, along with indigenous production indicates our crude oil consumption for the period 2011–12.

For the information of readers, TAPI is Turkmenistan - Afghanistan - Pakistan - India gas pipeline project.

4. Future Energy Scenario

4.1. Demand for energy

If we are not improving our population policy, by the year 2050, the population of India could be around 190 Crore, and at 2100, the same could be around 228 Crore. It is noteworthy that calculations have been made by assuming a steep decline in population from the current rate of rise of 18.6 per cent over a decade to 0 per cent at 2100. This prediction is made with the assumption that various energy forms are available to quantities demanded by the economy of the times in the coming decades.

Therefore, when the energy utilisation by economically advanced countries is likely to remain more or less steady in the coming times, the less developed economies including that of ours will demand energy at higher and higher levels for many more decades to come. This is because it becomes essential that the existing as well as the newly added population of ours be provided with better living conditions.

It is predicted that the demand for commercial energy forms in our country consisting of electricity, solid/liquid/gaseous fuels and non-commercial energy forms will rise to around 2 - 3 fold by the year 2050 and 3 - 5 fold by 2100.

By the year 2050, the demand for electricity in our country at 2 times the present level could be around 347,200 MW. At 3 times the present level the same could be around 520,800 MW.

4.2. Availability

4.2.1. Electricity

4.2.1.1. Electricity generation from green energy sources

In chapters on electricity generation from green energy sources, we had the assessment on the scope of these meeting our energy needs primarily for electricity generation. The following are the outcome:

Green energy sources in use

The limits of generating capacities of plants working on green energy sources could be as given below:

(1) Hydroelectricity – 60,000 MW.

(2) Wind electricity - 60,000 MW, de - rated to 25,000 MW of thermoelectric/hydroelectric/nuclear electric equivalent.

(3) Fission based nuclear electricity – 30,000 MW. (This quantum is included by taking an optimistic view on the probable capacity additions. However, a lot of uncertainties are seen to exist which could be adverse to reaching the target.

New green energy sources believed as promising

- *Our analysis of new green energy sources, namely, solar energy, geothermal energy and ocean wave/tide energy with respect to utilising them for electricity generation in a viable manner indicates that significant quantum of installed capacity is not likely to be available from these.*

- *We have also seen that for including energy from 'fusion based nuclear reactors' as a candidate for the future, we have to necessarily wait until a viable unit of the same has been developed and demonstrated. This is due to the extremely high level of complications involved in bringing a unit into reality.*

- *The possibility of change over to hydrogen economy which was thought as possible sometime ago, is not visible as of now, since, the related developmental efforts seem to have met with many road blocks.*

Considering all the above factors, we could conclude that the quantum of electricity which is likely to be available in our country from green energy sources in the long term could be in the range of 115,000 MW only. During the coming decades, the installed capacity could grow to this maximum value and then continue to be available at the same level.

4.2.1.2. Electricity generation from non-green energy sources

The difference between the demand for electricity and the extent of its availability from green energy sources is to be met from thermoelectricity generation. At a level of demand which is 2 times the current level, the installed capacity of thermoelectric plants is to be around 232,200 MW (this is twice the present capacity). Assuming that the demand at 2100 is 3 times the present level, the installed capacity of thermoelectric plants need be around 415,800 MW (this is around 3.6 times the present capacity).

The following points may be noted with respect to thermoelectricity generation:

- The thermoelectricity generating capacities in demand will be lower to the extent of energy intensity improvements achieved with respect to these plants. Units being established presently could be providing superior performance levels compared to the ones already in operation.

- Large scale capacity additions of coal based thermoelectric plants is to be brought in only by closing our eyes to global warming related issues.

- Large scale use of thermoelectric plants working on gaseous petroleum could address the environment related issues partially.

Considering the above factors we could conclude that the availability of electricity from coal based and natural gas based thermoelectric plants will be limited by the extent of availability of these fuels from time to time.

The above analysis clearly indicates that the role of green energy sources in the area of electricity generation in our country could come down significantly as a percentage of the total electricity generated in the coming times.

4.2.2 Fuels

4.2.2.1. Coal

Data on a few countries having higher level of coal deposits is given in the following table.

Table

Country	*Reserves (million tonne)*	*Global share (%)*	*Production – 2011 (million tonne)*
U.S.A.	246,643	27.1	1,004
Russia	157,010	17.3	334
China	114,500	12.6	3,576
India	92,445	10.2	586
Australia	78,500	8.6	414
South Africa	48,750	5.4	253
Ukraine	34,153	3.8	-
Kazakhistan	31,279	3.4	117
World – total	**909,064**	-	**7695.4**

(The above data is taken from Internet)

Note:

- The reserves of coal are believed to outlast that of petroleum.
- According to the current level of consumption, reserves of coal as indicated above is expected to last for about twelve decades only. However, considering the higher level of consumption in the coming decades, especially in the petroleum depleted era, reserves of coal might last only for a shorter time interval.
- India comes fourth with respect to the quantum of reserves which is indicated as 10.2 per cent.

- Our country experiences difficulty in providing coal as required for fuelling thermal power plants now itself. At present our requirements are largely met from indigenous sources.

- The related reports indicate that our deposits of coal largely are of low calorific value compared to imported coal (Ref: Report in the *Hindu Business Line* of November 8, 2009 titled 'Generation at key NTPC stations faulters').

In addition, the following factors put question mark on the availability of coal in our country according to its demand from time to time even in the near future.

They are:

- The recent media reports indicate that in order to meet the higher levels of demand for electricity of the coming decades, we are augmenting the installed capacity of thermoelectric power plants at high rate. Some of these power plants are proposed to be largely fuelled by imported coal. It is also reported that currently we are facing difficulties in meeting our projected demand for fuelling the above through import.

- The adoption of coal liquefaction process is considered essential to meet our ever increasing demand for fuels for the transportation industry particularly in the petroleum depleted era. This will further boost the demand for coal.

- A recent media report indicated that we are experiencing dificulties with respect to the opening up of new coal mines for want of environmental clearances. In addition, reports also indicate that new railway lines need be established for evacuating coal from newly opened mines. All these indicate that stabilising higher level of coal supply matching to the higher demand put forward by the newly built plants is a time consuming process.

- Global warming related issues currently being discussed in international forums are against utilisation of coal in a big way in the coming times.

The picture as projected above indicates that we in our country can not assume unlimited availability of coal for the future. Also, since our backbone for availability of energy in the form of electricity and fuels is coal, the possibility of a smooth transition of our economy to utilisation of energy sources other than petroleum in the petroleum depleted era is not visible.

4.2.2.2. Liquid petroleum

Data on crude oil reserves is given below:

Table

Country	*Proven reserves (billion barrels)*	*Production/year (billion barrels)*
Saudi Arabia	262.7	
Canada	178.9	
Iran	133.3	
Iraq	112.5	
UAE	97.8	
Kuwait	96.5	
Venezuela	75.9	
Russia	69.00	
India	5.7	
World – total	**1349**	**31.0**

(The above data is taken from internet)

Note: According to the data given in the above table, world reserves is 1349 billion barrels, while the yearly consumption is around 31 billion barrels.

Therefore the longevity of liquid petroleum is

$$= \frac{1349 \text{ bbl}}{31 \text{ bbl/year}} = 40 \text{ year}$$

It is well known that liquid petroleum has been in use for over a century; however, the use of natural gas in our country is comparatively of recent origin. Under these circumstances, shortages if any experienced in the availability of liquid petroleum during the period upto 2050 may be compensated by the use of natural gas to the extent the same is technically feasible.

In addition, the demand for liquid petroleum is expected to be on the rise during the coming decades and the new discoveries could compensate for the same. However, it is possible that the availability of liquid petroleum declines any time after 2050 and the same could bring in very serious repurcussions on the sustainability of our economy.

The above data also indicates that we Indians constituting around 17 per cent of the world population as of now, is blessed with only around 0.42 per cent of world liquid petroleum reserves.

4.2.2.3. Gaseous petroleum

(1) Data on natural gas reserves is given below.

Country	*Proven reserves (trillion cubic metre)*	*Production (trillion cubic metre)*
Russia	55	
Iran	33.5	
Turkmanistan	26.2	
Quatar	15.47	
U.S.A.	9.0	
Saudi Arabia	8.2	
Venezuela	5.524	
Nigeria	5.246	
Algeria	4.502	
China	3.100	
India	1.075	
World total	**300**	**3.359**

(The above data is taken from internet)

(1) Data on shale gas reserves is given below:

Country	*Proven reserves (trillion cubic metre)*
China	36.1
U.S.A.	24.41
Argentina	21.92
Mexico	19.28
South Afica	13.73
Australia	11.21
Canada	10.99
Libya	8.21
Algeria	6.54
Brazil	6.40
India	-
World total	**188**

(This data is from Internet)

The use of natural gas has picked up in our country later to that of liquid petroleum. In order that pollution due to smoke emitted by vehicles in metroes is controlled, we are gradually switching over to the use of natural gas to the extent feasible. In addition, this is preferred as a fuel for thermoelectric plants as well as feedstock for the manufacture of fertilisers and certain other chemicals.

We are able to make available liquid petroleum as required for internal consumption through import. At the same time, we are not able to ensure the availability of natural gas as required for internal consumption in the same manner. The reasons for the same include the following:

- Natural gas is transported efficiently to far off places through dedicated pipe lines. Our efforts to bring gas from countries in the West as well as in the East did not materialise largely due to issues including political ones.

- The alternative is to transport the same in dedicated ships provided with cooling arrangements to cool and liquefy the gas. Presently we are importing by adopting this route.

The situation with respect to availability of natural gas in our country is that we are not able to meet the present day demand. This could be understood from a news item that appeared in *The Hindu* of July 16, 2012 under the title "Nutrient based subsidy for urea a must". The report quotes a Planning Commission report and says that investments in urea production worth Rs. 25,000 to 30,000 crore are on hold because of lack of clarity on the availability of natural gas in the long term. The report adds that around 20 per cent of the existing urea capacity still operates on either naphtha or fuel oil as feedstock due to non-availability of natural gas.

4.2.2.4. Decline in the availability of petroleum

It is natural to expect a decline on the availability of liquid petroleum some time later to 2050. In anticipation of the shortage which might occur with regard to liquid fuels as mentioned above, it would be wise to make alternative arrangements before hand in the following manner.

Considering the fact that most of the liquid fuels are used for road transportation, we could shift a portion of road transportation to electric traction by installing high voltage overhead electricity supply lines along highways. This changeover will further increase the demand for additional installed capacities of thermoelectric plants, and in turn, the demand for coal. Assuring improved availability of coal is a must for exercising this option.

- The severe shortages in the availability of liquid as well as gaseous petroleum might kindle fresh interest in the utilisation of fission based nuclear electric plants and accordingly, they might contribute for the generation of electricity to the extent we have been able to create installed capacity in this area.

- The probable attainment of maturity of breeder reactors might enable generation of electricity from this green energy source in many countries. In case this energy source graduates to a most sought after one all over the world, there could be supply related issues with respect to availability of power plants as well as the timely replenishment of fuel. However, breeder reactors are likely to be a ray of hope beyond 2050.

- It might also be worthwhile to consider the objections that are being raised against the establishment of nuclear electricity generating parks, and accordingly, modify our plan to go in for large units. Perhaps we could change our policy in such a way that we establish smaller capacity units in a widely spread manner so that the fear among the general public is allayed. This could also help us to establish small units of Fast Breeder Reactors in a large scale as and when the same is ready for commercial use.

- Standardisation of coal liquefaction plants for industrial level production before 2050 could enable us to utilise them in a big way in the petroleum depleted era which is likely to come in later to 2050.

4.2.2.5. Availability of liquid fuels from renewable sources

The one and only one source of liquid fuels belonging to renewable energy category is fuels of biological origin. In the chapter on bio fuels, we had detailed discussions on the sources of these and the possibility of their coming to our rescue in a major way in the hydrocarbon depleted era.

The discussions have exposed the fact that the scope for bio fuel production in an effective manner is only with those countries blessed with large areas of cultivable land which could be earmarked for sugar cane cultivation (after domestic needs for sugar are met with). Brazil is an exception than rule.

This again clearly indicates the limited role of green energy sources for meeting the demand for liquid fuels in the hydrocarbon depleted era.

5. Long term energy scenario and its impact on economy and human population

The impact of lower level of availability of fuels in liquid/ gaseous forms and the limitations that exist in the availability of electricity and bio fuels from green energy sources is likely to result in the following scenario with respect to our economy in the long term.

It is difficult to think of air transportation without liquid fuels being available. On depletion of petrol, diesel and aviation kerosene, the availability of liquid fuels could be from the following sources:

- Coal liquefaction plants are built in a large scale all over the world. Production of hydrocarbons from tar sands and shale oil could become a common phenomenon. However, their commercial exploitation in a large scale and in turn their availability, will depend on proving their viability for production in a large scale as well as addressing issues related to environmental pollution in a satisfactory manner.

- The stocks of natural gas to the extent available cater to the numerous applications liquid fuels have been catering to until then.

- Since we are currently expanding the use of gaseous form of petroleum, the possible scenario is that the availability of energy from this energy source could come down significantly some time between 2050 – 2100. As a result of the reduced availability of fuels as indicated above, we could have the following scenarios:

- We have an air transportation sector operating at reduced capacities, and to that extent the so called global economy of the present times could reduce to regional/subregional economies.

- We have a surface transportation sector operating at a fraction of capacity of the previous times.
- We have an economy which is supported by different forms of energy at a much lower level, compared to the one which existed previously.
- The reduced availability of fuels and other forms of energy for supporting the then existing economy takes a heavy toll on the then existing human and animal population.

Adopting a realistic planning of the energy production and utilisation scenario from now on could reduce the intensity of a catastrophe taking place on the then existing economy, and in turn, on the population.

11

Some More Food for Thought on Energy Related Matters

1. Heightened availability of energy has caused accelerated rise in population in the twentieth century; and the same continues.......

The availability of energy and the human population existing at any point of time are closely related. Energy – Population planning is a single subject and it need be addressed accordingly. This is especially so in the present times, where in, we are fuelling our economy in a large scale using the non-renewable hydrocarbon based fuels.

This aspect is clearly visible in the data given below:

World population in billions (100 Crores)	*Year*	*Number of years it took for the increase of each billion (each100 Crore)*
1	1804	-
2	1927	123
3	1959	32
4	1974	15

(Contd.)

(Contd. Table)

World population in billions (100 Crores)	*Year*	*Number of years it took for the increase of each billion (each100 Crore)*
5	1987	13
6	1998	11
7	2011	13

Source: Report in *The Hindu* of October 17, 2011 titled, "World population to hit seven billion; severe stress on food, water and jobs".

It is crystal clear from the above table that the high rate of rise in our population which began in early twentieth century is man made, since, this coincides with the period during which we have started using coal and petroleum in a large scale.

By venturing into supporting the present day economy using the non-renewable energy forms of coal and petroleum in a large scale as mentioned above, we have indirectly taken the responsibility for its maintenance and upkeep into our hands, however large it might be and whatever the energy scenario it might be through the coming centuries. Along with the same, we have also taken the responsibility of meeting the day to day needs of the ever increasing population, whose strength has been indicated above. Acquiring this fundamental awareness is the core of Energy – Population plannir.g

2. We profess sustainable development; but....

Planners across the world are vouching that they are working towards sustainable development of our economies, however, it is doubtful that the same is practiced. This lack of planning could be seen from the manner in which we in India turn a blind eye to the high rate of rise in population, and also the manner in which we are depleting the reserves of coal and petroleum in a short time span of two centuries or so. We seem to be self-centred to the extent that we do not seem to have concern on welfare of future generations who are destined to live facing severe shortages of the energy forms which we are enjoying at present. It is essential to be aware, that we humans are responsible for their birth to the numbers likely to exist then, since, we have created an environment in which high quantum of energy is supplied in the form of coal and petroleum and this is a main factor for this increase in population.

3. We look after short term interests

It is not appropriate to point fingers on any person or any organisation for managing our economy in this way. Perhaps the world has always seen short term interests ruling the minds of its population, than ensuring welfare of ours in the long term. It is for the first time in the history of mankind that humans enjoying an economy which is serviced with unlimited availability of non-renewable energy in the form of coal and petroleum, has progressed in science, technology, inventions and innovations. We are watching this change with great disbelief, this is a fact.

4. Our ingenuity in creating innovative systems of the present is seldom used for the intelligent management of energy related issues

We do not seem to use the ingenuities that we used in the creation of today's unbelievably changed world in understanding the implications of depletion of the non-renewable liquid petroleum deposits. Issues concerning the long term availability or non-availability of liquid petroleum are generally not seen to be a topic of discussion among us. Prophecies/considered opinions on the possible short supply of crude oil have been made time and again, and they have all passed by in the same manner as prophecies made frequently on catastrophic ending of the world.

5. Objective evaluation does not seem to exist on energy availability in a long term

The Hindu of June 13, 2008 published an article titled, "World not running out of oil; says BP". Its highlight was a statement by the Chief of British Petroleum, wherein, he mentioned that the known deposits of crude oil are likely to support our economy only for about four decades more. He also pointed out that supply could be ensured only if permission is provided for mining the currently forbidden areas. As usual, we do not know as to where the reality lies. We also do not seem to have noticed this statement with the seriousness it deserves.

However, even if some one did react (the statement is a pointer to an extremely dangerous situation) it did not become a matter for public debate.

The *Malayala Manorama* daily of October 13, 2012 carried a news item titled, "Commissioning of the first unit of nuclear power plant at Koodamkulam soon". One of the highlights of the report was the statement made by the Chairman, Atomic Energy Commission that in 50 years or so the stock of hydrocarbons is expected to be depleted.

We have to see as to whether anyone takes note of this statement.

6. Projection of issues related to future energy scenario

Under these circumstances, it is appropriate to ask the following questions on the longevity of liquid petroleum deposits, the after effects of our depleting them off in the not so distant future and find convincing answers.

- Also taking into consideration the continuing rise in demand for liquid petroleum, how long its supplies are likely to be available to our asking; say for 50 years; 100 years?

- Under the present circumstances do we perceive that we have no method for assessing the longevity of liquid petroleum deposits? Does our responsibility end with using the same in a mad way and finishing it off?

- When the decline of liquid petroleum is it likely to begin, and how long the same will be available at reduced quantities?

- How could we manage supplies of liquid petroleum distillates during the period of their decline?

- Which are the key activities of our economy we will be compelled to cut down during and after the period of decline in the availability of liquid petroleum?

- Do we have sources for liquid fuels other than petroleum? In case we have alternatives, to what extent we shall be able to manage its supplies meeting the then demand, as well as taking care of the environment related issues?

- What will happen to our population in times of depletion of liquid petroleum distillates? In case we are not able to feed the then existing economy with liquid fuels to the then demand, how the transition to significantly lower level of utilisation of energy, and therefore lower level of population will be managed then?

- How we can prevent a catastrophic end in which bulk of human population vanishes from earth due to significantly reduced availability of energy ?

7. Back ground for projecting the above seemingly unaddressed issues

The published literature as available today do not seem to address such issues. On the contrary, they seem to provide a picture of confidence or over confidence. The following is the background for making this observation:

- As of now, liquid petroleum fractions are distributed in quantities at any one's asking; or we could say that there is no restriction on their consumption (eventhough the energy managers of today make statements that their reserves could vanish in 4 – 5 decades time). This situation could give the general public (*aam admi*) a feel that liquid petroleum is available in infinite quantities and for an infinite period; They are saying now: let us enjoy yarr!

- For the same reason, we (*aam admi*) do not recognise a need for restricting our consumption of petroleum or electricity for necessities alone. The manner in which gas guzzlers (luxury cars) are manufactured, distributed and lured the public to buy them (in India too) points out to the fact that we are creating and managing a vicious circle of high demand and meeting the same through higher level of supplies of petroleum distillates. Even the so called global warming related issues do not make us to think differently.

- One of the important assumptions made in the book titled, "Global Energy Perspectives" published by the Cambridge University Press in the year 1998 is that resource availability in various forms of energy will not be a major constraint for the development of twenty-first century economy. By the same, what the authors have assumed could be that no one seems to envisage a catastrophe like drastically reduced availability of energy during the twenty-first century. At the same time, the book does not give assurances on the continued availability of liquid petroleum to quantities demanded by us althrough this period. Also, the same does not provide similar assurances for twenty-second century and beyond. The above referred book is written based on studies conducted by the International Institute for Applied Systems Analysis (IIASA) and the World Energy Council (WEC). The principal analysis covers through 2020 and 2050 with some extensions to 2100.

- The published literature in the area of energy largely vouches that future is going to be one of sumptuous availability of energy from renewable energy sources. They propagate that solar energy and hydrogen are the energy forms that are going to rule the world in the coming times. We (*aam admi*) also get a feel that these energy sources could be developed to a scale that the heightened requirements of the coming times could be met by harnessing solar energy and hydrogen in a large scale.

- However, the viability concept referred here in our studies as a fundamental criteria for harnessing energy from weak energy sources such as solar, geothermal, ocean wave/tide has not been addressed in the above referred book. The possibility or otherwise of the fructification of fusion based nuclear reactors too has not been seriously debated. In some of the energy scenarios described in the book, a high rate of growth in the availability of energy from green and new green energy sources have been considered as a possibility.

- The above referred book refers to scenarios in which non-conventional, green energy sources are expected to substitute the conventional hydrocarbon based fuels in a big way, however, the expectations seems to have been based more on perceptions, than on evaluation.

Our progress in science and technology has made the world look very different from what it was about a century ago, this is for sure. As of now, a significant percentage of human population is living in comfort conditions that were unimaginable earlier. However, there looks to be negative impacts too. According to this author, this is visible in the form of blind beliefs in some of the areas concerning our day to day living.

8. Blind beliefs seem to rule the assessment on future energy scenario

Look at the following: We are always stressing on the adoption of sustainable development of our economy, and believe that our present day actions are in tune with this goal. However, we seem to lack clarity on how to ensure the same in the long run. This is evident from the manner in which we deplete the reserves of hydrocarbons. We are finishing off their stock, without ensuring that in the hydrocarbon depleted era, we get energy in quantities matching to the demand of the then existing economy from other energy sources. Probably, we blindly believe that our scientific and technological background is enough and more to derive huge quantum of energy in the forms we desire from the different energy sources that we perceive to be available all around us.

Our studies on the future energy scenario clearly exposes this blind belief, and the fact that we are not serious with respect to ensuring availability of various energy forms we are utilising and finishing off today, for the future generations.

Reports appearing frequently in the media regarding the establishment and utilisation of grid compatible solar electricity generating units in desert regions in a large scale are examples of the manner in which the related issues are perceived by us. The proposals envisage that solar electricity generating plants established in Sub Sahara regions are of immense value with respect to generating and providing grid compatible electricity to Europe (This statement is based on newspaper reports).

It is also generally said (according to newspaper reports) by energy administrators of the developed world that developing countries like India need to focus on harnessing solar energy for a variety of applications rather than looking for the use of more and more quantities of conventional fuel for quenching our thirst for energy. However, this author is of the opinion that if such proposals are evaluated from a viability angle, they would have been discarded in the initial stage itself.

There was another proposal regarding the establishment of solar electricity generating units in a big way in the desert regions of South Africa. The reports were highlighting that even though they were very costly propositions, once set up, the units will be of immense use to the country wherein a wide gap exists between the demand and supply of electricity.

The fact is that no analysis seems to have been carried out on the reasons behind the unacceptable levels of high cost of solar electricity generation, such as its viability aspects.

9. Let us face it.... the alternatives are attractive, but not feasible

The Hindu of March 11, 2012 carried an article titled "Let us face it... the alternatives are attractive, but not feasible". This article was authored by Ipshit Tarun. The relevant points regarding the scope of alternative energy sources is given below:

- Renewable energy sources are attractive, but in a sense, powerless.
- Maybe, some day we will all live in houses with photo voltaic roof tiles, but in the real world, a 1000 MW of solar electric plant will require 60 square miles of solar panels.
- A few basic questions regarding harnessing electricity from various energy sources put forward are: (1) when demand rises, one can fire up more coal. By this, what he means is that thermoelectric plants can produce electricity for 24 hours a day if required. (2) At the same time how one will cause wind to blow of sun to shine 24x7?

The above comments are made in the background of the live issues we have been facing for the past many months regarding the commissioning of Koodamkulam Nuclear Power Plant of 2000 MW installed capacity.

This is the same prime issue raised by me in this book and based on which conclusions are drawn. The detailed analytical studies have brought out the fact that solar electricity generation is a non-viable process of electricity generation; and the same could be the case of electricity generation from sea waves/tides and geo-thermal heat which are considered by many as the unlimited sources of energy for the future. This is what makes the future energy scenario a dangerous one.

12

Executive Summary on Green Energy Sources and Future Energy Scenario

It is essential that we carryout a thoughtful and reasoned out analysis of long term energy scenario of the world due to the following:

1. **Relationship between energy use and the strength of human population need be taken note of**

1.1. Energy is essential for the evolution and existence of life

Life on our planet is supported by among others, oxygen, water and food. Energy from food is as important as the other two for the evolution and existence of life. In addition to the energy used for food production activities, we need energy for carrying out activities that enable a comfortable living. All these energy requirements put together constitute many orders per capita energy used centuries ago.

1.2. The strength of human population at any instant in the history of mankind and the quantum of energy used then are related

The strength of human population that existed at any point of time in the history of mankind and the quantum of energy we used then are related. Prior to 1700 AD, our economy was largely supported by green and renewable energy sources. For example sunlight provided illumination, firewood fuelled hearths, seasonal winds powered country boats, solar heat dried agricultural products and the muscle power of humans and draught animals moved goods.

The present day economy is like an inflated balloon which is fuelled by huge quantum of energy from coal and petroleum. These are non-renewable and non-green energy sources. Not only that the economy is like an inflated balloon and could burst on short supply of energy; the heightened use of energy sources has also resulted in rise of human population matching to the quantum of energy used. The large size of the present day economy as well as the human population are the result of increased utilisation of energy from non-renewable energy sources for over a century, and hence is man made.

2. Large scale use of petroleum

2.1. Unplanned and unrestricted use of petroleum

We are mining, processing and burning away liquid and gaseous petroleum in an unplanned as well as unrestricted way. Looking at the manner in which deposits of liquid and gaseous petroleum are being finished off seems to provide a false hope to us (*aam admi*) that they will be available to us until the universe exists. This also results in their wastage; the reserves of which would have otherwise lasted longer.

2.2. Prediction on shortage of liquid petroleum

By using liquid petroleum for the past one century in huge quantities, we have brought in a situation, wherein, its reserves have come down drastically. This was indicated by the Chief of British Petroleum (refer report in *The Hindu* of June 13, 2008 titled, "World is not running out of oil; says B.P"). The essence of what he mentioned is that liquid petroleum is expected to be available for around four more decades only. This serious warning seems to have gone unnoticed among us. No discussion seems to have taken place among energy managers/administrators on the possible imminent shortage of petroleum and its negative effects on the then existing economy. We have dismissed this warning on the same lines as we dismiss frequent warnings on ending of the world made off and on from different quarters.

Recently, a statement was made by the Chairman of Atomic Energy Commission; Dr. Anil Kakodkar related to the same issue. While talking on the Koodamkulam issue, he highlighed that hydrocarbons could be available to us only for around five decades more from now on. It is possible that he was mentioning about the possible short supply of petroleum after a few decades now and the need to pursue with the use of nuclear reactors.

2.3. Transportation modes are energised by liquid petroleum

Keeping aside a portion of railway traction, the transportation modes we use on land, sea and air are fuelled largely by liquid petroleum derivatives. Any shortage in their availability will cause our economy to burst like an inflated balloon, and change the world economy, into regional/subregional economies.

3. Importance of high intensity of energy sources in ensuring the viability of energy conversion processes

3.1. Non renewable energy sources currently in use and their high energy intensity

We produce fire by burning firewood, petroleum and coal, and utilise the thermal energy thus obtained for enabling domestic as well as industrial processes. The energy intensity of fire is very high. In comparison, the energy intensity of solar, wave/tide, geothermal energy sources is low. Most electricity is produced in our country from thermoelectric plants fuelled by coal. Nowadays, large scale capacity additions are undertaken in this area. However, shortage in the availability of coal could be a factor that is adverse to generation of unlimited quantum of electricity from these.

Hydroelectric and wind electric plants (these are renewable and green energy sources) provide electricity. Their combined installed capacity is around 20 per cent as of now. Capacity additions are limited by the extent of their respective potentials. Only around 3 per cent of our total electricity we use in India is generated from fission based nuclear electric plants (this is a non-renewable green energy source). In addition, we are facing many hurdles with respect to moving forward with the proposed accelerated capacity additions in this area.

3.2. We are looking for substitutes for fuels with energy sources belonging to green energy category

Considering the depletion of stock of petroleum which could come in during the second half of the twen'y-first century itself as indicated above, as well as the global warming related dangers which are believed to be at our door step, we are looking for alternative sources of energy, particularly belonging to green energy category. These new energy sources include solar energy, geothermal energy, wave/tide energy. We desire that these hitherto unused energy sources are harnessed in a large scale as electricity as well as thermal energy and are used in place of the presently used hydrocarbon based fuels.

3.3. New green energy sources are of low energy intensity

One common characteristic of the above mentioned hitherto less used energy sources is that compared to the energy intensity we get from fuels, the energy intensities of these sources are very low and therefore are weak energy sources. This is evident from the fact that we can expose ourself to hot sun for a long time, whereas we can not stand inside a fire.

Here in is the need to assess the viability aspects of the different energy sources with respect to the energy conversion processes used, for example; conversion of solar energy into electricity. This could also be called as energy balance characteristics.

4. Energy balance characteristics of energy conversion processes

- Production of electricity from various energy sources is by the adoption of energy conversion processes.

Examples are :

(a) Conversion of thermal energy obtained by burning of coal into electricity.

(b) Conversion of the kinetic energy of wind into electricity.

(c) Conversion of solar energy into electricity.

- Before we go ahead with electricity generation from an energy source in a big way, it is essential that we are clear about the energy balance criteria of the energy conversion process we utilise for the same.

- The concept of energy balance is defined as given below. For carrying out any activity, it is essential that we input energy into it. The same is the case with the manufacture and operation (for lifetime) of electricity generating plants too. Presently, the energy input into this activity is largely in the form of electricity, since, the related manufacturing activities are carried out using machines powered by electricity. At the same time, the output from the plant is the electricity generated during its life time operation. For a power plant to function as an electricity generating one, the output energy available to us as electricity during its life time operation has to be many times the input energy.

- The above said analysis has been carried out with respect to the different types of electricity generating power plants. The results of the analysis of the energy conversion processes related to the different types of electricity generating plants is as given below:

- Electricity generation from sources with high energy intensity such as fuels is found to meet the energy balance requirements. Hence these power plants are found to be viable electricity production units. By this, what we mean is that, the input energy for their manufacture and lifetime operation is a minor portion of the quantum of electricity available from them during their life. In other words, after discounting for the energy used for the manufacture and lifetime operation of the plants, significant quantum of electricity is available from them for meeting our varied uses in both industrial as well as domestic sectors (details are available in the chapter – 7, titled, 'pointers to viability of electricity generating power plants').

- It is also inferred that the processes of electricity generation from low intensity energy sources (e.g., solar, wave/tide, geothermal energies) do not meet the net energy analysis criteria. Out of these, the case of solar electricity generation has been analysed in detail. Based on the same, it is concluded that solar electricity generation methods using the currently available technologies are non-viable processes, since, the energy input for the manufacture and life time operation of the plant (this is largely in the form of electricity) is seen to be more that the electricity we get from the same during their lifetime operation.

- Details are given in chapter - 7 in a concise form. For ready reference this data is reproduced in the following table.

5. The need to study the viability of electricity generation from weak energy sources and bring out clarity on the related aspects.

- Our study on energy balance characteristic of solar electricity generation indicates that the same is a non-viable process and hence, solar photo voltaic and solar thermal electricity generation units do not perform as electricity generation plants. This is so particularly with respect to grid connected ones.

- This aspect of non-viability has not been brought out by professionals and economists working in the energy field so

Table: Pointers to Viability or Non-viability of Electricity Generating Plants

Power Plant	*Cost per unit of electricity generated (Rs.)*	*Number of similar plants producible from net energy*	*Expenses incurred for generating 157.68 million units over 30 year period (Rs. in millions)*	*Remarks*
Hydroelectric	1.63	11.7	258.5	Expenses are the lowest, hence most viable
Thermoelectric	5.27	2.93	832.7	Viable
Nuclear electric	4.67	3.4	738.1	Viable
Wind electric	4.52	3.47	712.8	Viable
Solar electric	25.5	Nil	4026.0	(1) Expenses are about 15 times the lowest. (2) No net energy produced, hence points to non-viability

far. Due to the same reason, installation of solar electricity generating plants is being pursued in a large scale in our country by providing heavy subsidies at all stages of solar electricity generation. However, we do not recognise many negative aspects of solar electricity generation; these are clearly brought out in this book.

- In the same lines we need to bring out clarity on the various aspects of electricity generation from other weak energy sources we consider to be promising for the future. Examples are wave/tide energies and geothermal energy.

- If we do not develop an understanding on all the related positive and negative aspects of electricity generation from weak energy sources at least now, we could be continuing to hold false hopes as given below:

    - Solar energy is the energy of tomorrow.

    - Solar, wave/tide, geothermal, wind energies will be able to provide unlimited quantum of energy in the form of electricity and heat in hydrocarbon depleted era.

6. **Need to have a thoughtful and reasoned out analysis of the future energy scenario is important for sustaining the future generations as well as economy due to the following:**

- An objective analysis related to the viability of electricity generation processes is not usually found in the related techno-economic writings.

- The world population is continuously on the rise. This rise is largely in developing countries like that of ours, and is enabled by the use of coal and petroleum in unbelievably huge quantities.

- Due to many factors including the high rate of rise in population of the present, the energy needs of our economy is also continuously on the rise. However, since the reserves of liquid petroleum are depleting and are not believed to last long, the economy of the times is likely to face the danger of break down. This will have serious repercussions on the then existing populations also.

- All these point to the need for a realistic assessment of the future energy scenario and the need for the Energy – population – Economy planning of the coming times accordingly.

- Energy – population – Economy planning is a single subject, and therefore, any assessment which is away from the same will result in catastrophic end of future generations.

Annexures

1. Nuclear Fuels

(A) Fuels for nuclear fission

Otto Hahn in 1939 discovered a very important reaction while studying the effect of bombardment of fast neutrons on uranium. The products of the reaction were Barium and Krypton. In addition to these products, a huge amount of energy was liberated. The reaction could be represented as:

$$U^{235}_{92} + n^{1}_{0} = Ba^{144}_{56} + Kr^{90}_{36} + 2\,n^{1}_{0} + \text{energy}$$

The characteristics of the reaction are

(i) Uranium nucleus is split into two nearly equal parts.

(ii) A considerable amount of energy is released.

(iii) Process is a self-sustaining chain reaction, since for every neutron causing the fission reaction, two to three neutrons are generated.

Thus, nuclear fission could be described as the disintegration process in which, a heavy nucleus on capturing neutron, splits into approximately two equal parts, with the liberation of energy.

Further studies have brought out the following information regarding the various related reactions.

(i) $U^{235}{}_{92} + n^{1}{}_{0} = Ba^{144}{}_{56} = Kr^{90}{}_{36} = 2n^{1}{}_{0}$ + energy

(slow or fast)

(ii) $U^{238}{}_{92} + n^{1}{}_{0} = U^{239}{}_{92}$ + gamma rays

$U^{239}{}_{92} = Np^{239}{}_{93} + e^{0}{}_{-1}$ (beta particle)

$Np^{239}{}_{93} = Pu^{239}{}_{94} + e^{0}{}_{-1}$ (beta particle)

(iii) $U^{238}{}_{92} + n^{1}{}_{0} = Xe^{143}{}_{54} + Sr^{94}{}_{38} + 2n^{1}{}_{0}$ + energy

(This is not a common reaction)

(iv) During the fission of $U^{238}{}_{92}$ by slow neutrons, a mass change of 0.207 amu (atomic mass units) takes place, and an equivalent energy of 200 MeV is released.

(v) For the complete fission of $U^{235}{}_{92}$ atoms contained in 1 kg, the release of energy would be about 20 million kWh. This is equal to the energy released by burning 3000 tonne of coal.

(vi) Fission reaction is a self propagating chain reaction.

(vii) Every released secondary neutron in uranium fission does not strike a nucleus and cause further fission. Some neutrons miss their target and escape into air and are lost. For maintaining fission reaction in $U^{235}{}_{92}$, it must be larger than critical mass, otherwise, too many secondary neutrons escape into the air, and hence, chain reaction can not be sustained. On the other hand, relatively fewer neutrons escape from 'super critical mass', and as a result, a chain reaction is set up. Thus, subcritical size is safe, supercritical size needs control.

Fuels used in a nuclear reactor are:

Natural uranium (99.28% U^{235}_{92} + 0.714% U^{238}_{92})

Enriched uranium (contains higher per cent of U^{235}_{92}).

U^{233}_{92} obtained from Th^{232}_{90}

Pu^{239}_{94}, obtained from U^{238}_{92}.

Uranium is the fuel for fission based nuclear reactors. India has very less proven deposits of uranium ore.

Thorium is an element that could be converted into fissionable isotope of uranium by manufacturing it as part of fuel pellets burned in nuclear reactors. It is extracted from Monazite sand; large deposits are found in India. Monazite contains 1 to 10 per cent of ThO_2. Once we become successful in the realisation of breeder reactors on a commercial scale, we shall be able to utilise our vast thorium deposits for the generation of power.

(B) Fuels considered for use in nuclear fusion reactors

Fusion of protons is the most difficult operation. It is known, however, that nuclei of heavier isotopes of hydrogen (Deuterium and Tritium) fuse more readily than protons. Examples of such fusion reactions are:

$$D^2_1 + D^2_2 = He^3_2 + n^1_0 + 3.3 \text{ MeV (Deuterium)}$$

$$D^2_1 + D^2_1 = T^3_1 + H^1_1 + 4.0 \text{ Me (Tritium)}$$

$$D^2_1 + T^3_1 = He^4_2 + n^1_0 + 17.6 \text{ MeV}$$

$$H^1_1 + T^3_1 = He^4_2 + 20 \text{ MeV}$$

$$T^3_1 + T^3_1 = He^4_2 + 2n^1_0 + 11 \text{ MeV}$$

$$D^2_1 + D^2_1 = He^4_2 + 24 \text{ MeV}$$

Tritium has to be manufactured by neutron bombardment on Lithium-6.

$Li^6_3 + n^1_0 = He^4_2 + T^3_1$.

However, this tritium is not stable.

If the reaction between two deuterium nuclei could take place completely by their fusion, 1 kg of deuterium would provide us as much energy as 2.865 million kg of coal.

2. Net Energy Analysis

Awareness on the need for carrying out 'net energy analysis' with respect to energy conversion processes we intend to adopt afresh, seems to be lacking among the present day energy professionals and administrators. The related literature rarely highlights this aspect; perhaps the same could be the reason for this study not getting attention nowadays.

Proposals for the wide spread adoption of electricity generation from weak green energy sources seem to be order of the day. This is particularly visible with respect to the establishment of grid connected solar electricity generating plants in our country. Solar photo voltaic plants hitherto used for special applications only until the recent past has been chosen to be established in a large scale for grid connected applications by providing heavy subsidies at all stages.

However, detailed analytical studies carried by this author and presented in this book convincingly indicate that net energy available from these is negative and therefore this author wishes that independent studies are carried out and this aspect is confirmed before large sum of money is invested in solar electricity generating plants.

A brief on a net energy analysis study carried out on a nuclear power plant by USA is given below. The study was carried out to provide a guidance for selection of energy sources for different applications – this is clearly indicated in the same. The source for his information is an Energy Handbook published over three decades ago.

Study

Net energy analysis refers to studies on the amount of electricity generated by a power plant relative to the amount of external energy required to construct, fuel and operate the plant. The Energy Research and Development Administration has made such studies for light water reactor nuclear power plants.[1] Drawing from the studies by the State of Oregon, the Centre for Advanced Computation of the University of Illinos (Pilati and Richard), Development Sciences, inc., and the Institute of Energy Analysis, ERDA has drawn the conclusion that light water nuclear power plants produce substantially more energy than is required to construct, fuel and operate the plants. The major conclusions are :

- The direct and indirect external energy inputs are about 26 per cent of the energy output. This value is within the values for electrical engineering facilities fuelled with conventional fossil fuels.

- Uranium enrichment, using the diffusion process, accounts for about 91 per cent of all direct energy inputs and 72 per cent of all direct and indirect energy inputs.Use of gas centrifuge now being developed would significantly reduce the energy required for enrichment. The construction and operation of the power plant accounts for another 16 per cent.

- The internal energy loss due to the thermodynamic efficiency of a nuclear plant is about 68 per cent. This means that the nuclear fuel must generate 3000 Btu of energy in order to produce 1000 Btu of output as electricity.

- According to the 1975 studies (Development Sciences, Inc., and Oregon), 6000 to 7000 Btu of resource base must be available in order to have the 3000 Btu energy input to the reactor. This additional resource base is necessary to allow uranium not recovered during mining or not during processing as well as fissionable uranium left in the tails during the enrichment process. However, all the 3000 to 4000 Btu of energy remaining are not irrevocably lost or consumed; much of the energy may be utilised in future if the economics of processing these resources become attractive.

1. Appendix B, A National Plan for Energy Research, Development and Demonstration: Creating energy choices for the future, ERDA 76 – 1, Energy Research and Development Administration, April 15; 1976.

Table: Comparison of Net Energy Results

Investigator	*Units of External Energy Input per 1000 units of Output*
Development Sciences, inc.	238
State of Oregon Study	194
University of illinos, Centre for advanced computation (Pilati and Richard)	210
Institute of Energy Analysis	248
ERDA 76 – 1	262

Source: Appendix B, a National Plan for Energy Research, Development and Demonstration: Creating Energy choices for the Future, ERDA 76 – 1, Energy Research and Development Administration, April 15, 1976.

Table: Energy Requirements for a Large Nuclear Power Plant

1000 MW Pressurized water Reactor – No Recycle – 0.20% Enrichment Talls assembly – 30 year Life – 61% average Capacity Factor – 160,3000,000 MWh Output (547 Trillion Btu)*

Electrical inputs, in MWh

Process	*Quantity*	*Direct*	*Materials*	*Construction*
	3909 MTU	47,760	22,050	6,210
Mining	3909 MTU	67,430	16,140	2,780
Milling	3909 MTU	39,830	16,620	620
Conversion	3124 X 10^3 SWU	8,778,000	18,120	24,050
Enrichment				
	683 MTU			
Fuel fabrication		67,750	137,200	600
Power plant				
Construction and	30 Years			
operation	683 MTU	0	256,500	205,000
Fuel storage	30 Years	7,280	2,430	3,880
Waste storage		130	4,560	320
Transportation	3909 MTU			
- Natural	693 MTU	0	410	0
- Fuel		0	1,546	0
Total		**9,008,000**	**475,600**	**243,40**

Thermal inputs, in million Btu

Process	*Direct*	*Materials*	*Construction*	*Total inputs, Billion Btu* **
Mining	1,005,000	621,500	441,500	2,935
Milling	1,310,000	621500	121,200	3,037
Conversion	4,826,000	426,000	29,000	5,334
Enrichment	1,048,000	428,000	1,016,400	103,037
Fuel fabrication				
Power plant	147,900	1,579,00	26,000	4,096
Construction and operation				
Fuel storage	378,000	8,179,000	9,583,000	23,401
Waste storage	4,560	64,300	174,300	240
Transportation	2,080	158,200	22,950	398
- Natural				
- Fuel	31,900	24,470	0	61
	120,300	92,290	0	230
Total	**8,874,000**	**12,195,000**	**1,1414,000**	**142,769**

*Electricity converted to thermal equivalent at 3413 Btu per kilo watt hour.
Legend; MTU – Metric Tonnes uranium.
SWU – Separate work units (directly proportional to the energy required for the enrichment process)
**Electricity converted to thermal equivalent at average heat rate of 11,400 btu per kilo watt hour.

Note: The conversion of electrical to thermal units plus the addition of thermal inputs from different energy sources ignores the quality aspect of preferred fuels. This is a serious shortcoming as there is no single factor that is satisfactory for converting different energy sources to an equivalent base.

Source: Institute for Energy Analysis with adjustments by ERDA.

Source: Appendix B, A National Plan for Energy Research, Development and demonstration: Creating Energy Choices for the Future, ERDA 76 – 1. Energy Research and Development Administration, April 15, 1976.

Index

T

U

V

W